ADVANCE PRAISE

START YOUR FARM

"*Start Your Farm* is an eminently readable, entertaining, and important book for new farmers. The authors don't hold back on the tough parts of farming. They coach the reader on how to overcome obstacles and reap the financial and emotional rewards on the other side. I recommend this book to anyone looking to make the brave leap into a farm career."
—**Lindsey Lusher Shute,** cofounder of the
National Young Farmers Coalition

"This book is full of wisdom, wit, and practical advice from two of the most respected figures in agriculture today. Anyone interested in farming for a career should pick up a copy. There is no career more rewarding than farming, and this book eases the burden for those starting out."
—**Ben Hartman,** author of *The Lean Farm* and
The Lean Farm Guide to Growing Vegetables

"A must-read for anyone who feels called to be a farmer. *Start Your Farm* is the most insightful, honest, challenging, and yet joyful and hopeful book I have read for would-be farmers. Prichard's and Polishuk's decades-long experiences in striving to farm sustainably also provide valuable insights into what it will take to farm successfully in the twenty-first century. Read this book and learn from two truly professional farmers."
—**John Ikerd,** professor emeritus of agricultural economics,
University of Missouri

"If you have romantic Instagram ideas about living off the land, you need to read this book. Jam-packed with practical advice and real-life vignettes, *Start Your Farm* gives you the down and dirty—the financial, emotional, and physical aspects of running a working farm. But this gritty realism is perfectly balanced with hope and encouragement. Pritchard and Polishuk not only talk the talk, but walk the walk, and they will guide you through the intricacies of starting a farm of your own. If you have what it takes, that is!"
—**Lisa Steele,** author of *Fresh Eggs Daily*

"If you dream of farming, *Start Your Farm* provides proven practical advice, delivered with a sense of humility and humor. It's as much a 'why' as a 'how-to' book, separating it from the plethora of guides for beginning farmers. Clear, concise, and highly useful, this book will help you make smarter decisions and fewer mistakes. I wish I had access to this kind of wisdom when I started farming forty-eight years ago. Get out there and farm, and carry this valuable info with you!"
—**Amigo Bob Cantisano,** founding member of Organic Ag Advisors and the Ecological Farming Association

START YOUR
FARM

Also by Forrest Pritchard

Gaining Ground:
A Story of Farmers' Markets, Local Food, and Saving the Family Farm

Growing Tomorrow:
A Farm-to-Table Journey in Photos and Recipes

START YOUR
FARM

THE AUTHORITATIVE GUIDE
TO BECOMING A SUSTAINABLE
21ST-CENTURY FARMER

FORREST PRITCHARD
& ELLEN POLISHUK

THE EXPERIMENT

NEW YORK

START YOUR FARM: *The Authoritative Guide to Becoming a Sustainable 21st-Century Farmer*
Copyright © 2018 by Forrest Pritchard and Ellen Polishuk

Excerpt of Wendell Berry's poem "The Contrariness of the Mad Farmer" used by permission of the author.

All rights reserved. Except for brief passages quoted in newspaper, magazine, radio, television, or online reviews, no portion of this book may be reproduced, distributed, or transmitted in any form or by any means, electronic or mechanical, including photocopying, recording, or information storage or retrieval system, without the prior written permission of the publisher.

The Experiment, LLC
220 East 23rd Street, Suite 600
New York, NY 10010-4658
theexperimentpublishing.com

THE EXPERIMENT and its colophon are registered trademarks of The Experiment, LLC. Many of the designations used by manufacturers and sellers to distinguish their products are claimed as trademarks. Where those designations appear in this book and The Experiment was aware of a trademark claim, the designations have been capitalized.

The Experiment's books are available at special discounts when purchased in bulk for premiums and sales promotions as well as for fund-raising or educational use. For details, contact us at info@theexperiment-publishing.com.

Library of Congress Cataloging-in-Publication Data

Names: Pritchard, Forrest, author. | Polishuk, Ellen, author.
Title: Start your farm : the authoritative guide to becoming a sustainable
   twenty-first-century farmer / Forrest Pritchard & Ellen Polishuk.
Description: New York, NY : The Experiment, LLC, [2018]
Identifiers: LCCN 2018024897 (print) | LCCN 2018028728 (ebook) | ISBN
   9781615195138 (ebook) | ISBN 9781615194896 (pbk.)
Subjects: LCSH: Sustainable agriculture--Handbooks, manuals, etc.
Classification: LCC S494.5.S86 (ebook) | LCC S494.5.S86 P757 2018 (print) |
   DDC 338.1--dc23
LC record available at https://lccn.loc.gov/2018024897

ISBN 978-1-61519-489-6
Ebook ISBN 978-1-61519-513-8

Cover by Becky Terhune
Text design by Sarah Smith
Cover photographs by Molly M. Peterson, except top center photograph © christian42 | Adobe Stock
Photograph of Ellen Polishuk by Diana Kolnick
Photograph of Forrest Pritchard by Molly M. Peterson

Manufactured in China

First printing September 2018
10 9 8 7 6 5

To my parents, who believed
—FORREST

To Hana, who kept opening doors for me, giving me the opportunity to soar
—ELLEN

"There is only one success . . . to be able to spend your life in your own way."
—CHRISTOPHER MORLEY

CONTENTS

INTRODUCTION

This will be unlike any farming book you have ever read.

If your mind is intent on solving problems, if your spirit yearns for the satisfaction of meaningful work and honest results, and if your body is ready for the arduous, physical toil that accompanies all agriculture, no matter what you eventually decide to grow—then the following pages should be useful to you, the aspiring sustainable farmer. At minimum, this book will lead to introspection and reflection, both important for farming success. It's our hope that it will offer a road map to a life of creativity, collaboration, and hard-won satisfaction. These are goals, in our opinion, worth pursuing.

Naturally, there are scores of instruction manuals, many of them excellent, that specifically address when to plant broccoli, how to build a chicken coop, or what amendments are required to balance soil pH. Additionally, there are entire bookstore sections devoted to starting a new business, the nuts and bolts of management, marketing, and finance, and titles encouraging readers to think outside the box. We have no intention of adding more of the same to the leaning stack of books that might be on your nightstand at this moment.

Instead, *Start Your Farm* is a hybrid of multiple seemingly diverse elements: agricultural insight, business acumen, and self-help wisdom. In biology, *heterosis*—the act of crossing genetically distinct progenitors to achieve a stronger, more resilient offspring—is a fundamental tenet. We will intentionally "cross" multiple concepts—economic, ecologic, and cultural—and cultivate something greater than these individual parts. In philosophy, this is known as gestalt theory, a mathematical oddity where 1 + 1 + 1 somehow ends up equaling 4. In farming, this is often called holistic agriculture, where the individual instruments of soil biology, seasonality, and human management blend into an ecologic (and occasionally economic) harmony. Don't be concerned if this sounds a little highfalutin at the moment. We'll spend lots of time explaining these philosophical concepts, using stories and examples to demonstrate why they're so valuable to understand.

Your authors come from very different agricultural backgrounds. Forrest grew up on a two-thousand-acre Shenandoah Valley farm that raised everything from corn and soybeans to apples and cherries to cattle, pigs, and chickens. As a young, seventh-generation farmer, he transitioned his family's production from mainstream commodity-based agriculture into an independent, direct-marketed sustainable livestock operation.

Ellen, on the other hand, was raised in the suburbs, and had no farming background at all. Her interest in agriculture blossomed during her teenage and early adult years, and it took shape when she was hired to manage a large vegetable farm in the early 1990s. With her help, the farm flourished into one of the most successful and admired operations in the mid-Atlantic.

With such strikingly different paths to agriculture, together we offer a wealth of farming experience to share. Thus, we have divided up the chapters to best reflect our respective passions and areas of expertise.

Still, a project of this scope has natural limitations, and you should be apprised of these challenges well in advance. Though your authors have each built successful farm businesses, enjoying high peaks and navigating the lowest swales of farming, and, combined, bring roughly fifty years of hard-won agricultural experience to assist you in your journey, let's be perfectly clear: We can't tell you *how* to farm. That would be akin to explaining through a book how to sing or paint or dance. After learning all the means and methods, there is always an element of artistry involved, and this shouldn't be undervalued. But even more important are a handful of unquantifiable components: your individual capacities, your personal circumstances, and all the heart you can possibly muster.

So how can we address these myriad variables in any sort of useful way? We believe it starts with being honest throughout. Sure, we can offer time-worn advice and useful generalizations, and we will. We can provide anecdotes about farming, business, and marketing, and we will. We can offer rural humor that will help mitigate the harsh life-and-death realities of agriculture, and we'd better!

But to make this book a truly useful tool, we must push into somewhat uncharted territory. This means suspending many core beliefs: beliefs in mainstream agricultural thinking, in conventional economic motivations, and in traditional life destinations. Call it virgin territory, or call it the wild. Regardless, we'll do our best to help you navigate this unfamiliar terrain.

Start Your Farm will lay the foundations that all good farmers possess: resilient problem-solving, sure-footed prioritization, calm analytical thinking, and devoted profit-making. Do you have it in you to transform five acres into a productive, financially viable farmstead? How about fifty, or even five hundred? By the end, we think that you will. And we will provide ideas and concepts that will be translatable, scalable, and revenue-generating.

Nevertheless, as farmers, we remain faced with problems beyond our control. In the twenty-first century, the price tag of land is more often commensurate with building a house, a subdivision, or a shopping mall than with agricultural use. This is especially true within a hundred-mile radius of any major American city, or within twenty-five miles of any midsize town. It's a departure from as recently as a few decades ago, when open land beyond city limits was much more commonly used for farming. Over the course of only two generations, finding affordable land has become nearly impossible for the new producer. This is a major issue of our time, and one we'll fully address.

Compounding this problem is that land prices are often so high that, for established farmers, especially those whose children or family don't want to take over the farm, selling their land often seems like their only viable retirement option. To help solve this, in *Start Your Farm*, we lay out multiple investment and retirement strategies that *don't* involve selling the farm. When farmers sell land, these acreages commonly end up as subdivisions or commercial centers, creating less land supply and more demand, driving prices ever higher. It's a vicious cycle, to be sure.

Thus, the majority of affordable land ends up being located far away from a concentrated customer base, limiting farmers' options as to how their goods can be marketed. All combined, these overlapping factors have created self-perpetuating momentum toward bigger and more specialized farms, with the vast majority of producers today (roughly 96 percent) growing only a narrow variety of commodities: corn, soybeans, wheat, hay, cotton, and livestock.

Challenging? Absolutely. But these are the facts of contemporary agriculture, realities that would-be farmers, and even established producers, must face. Economic headwinds this strong shouldn't be taken lightly, and we'll discuss how and why this "commodity system" operates. That said, each new challenge also creates a landscape of new possibilities, and it's our aim to point out these opportunities, navigating you toward improved odds of success.

Because land affordability (or lack thereof) is a primary concern, we will specifically focus on small- to medium-size acreages, plots ranging from one to two acres on the small end to perhaps five hundred acres on the high end. There are three main reasons for this.

First, acreages of this range will be far easier for a new farmer to capitalize or access, either through purchasing, leasing, or—best of all, as we'll explain later—creative acquisition. Second, for perhaps the first time in history, we believe that technology, accessibility to markets, and growing customer demand for sustainably produced food have all intersected in such a way that small- to medium-scale farming enterprises can now be economically viable. In fact, we believe these might be the *most* financially stable farms of the future. Finally, it is extremely difficult to avoid commodity or wholesale markets and their accompanying pricing constraints when farming on a large scale (in this case, meaning more than five hundred acres). We recommend having as much production, marketing, and financial flexibility as possible, which naturally lends itself to more modestly sized operations.

All of that said, *Start Your Farm* will certainly be of use to established farmers who currently own or operate larger farms. If you have five hundred or more open acres, it probably means that some type of farming operation already exists. The information we'll provide should be equally valuable to large-scale farmers looking to transition into a new agricultural enterprise, reevaluating how they do business, or preparing a younger generation to take over a preexisting operation with an eye toward the future.

We have considerable experience working on or with larger farms, and we understand how they operate. We recognize that accompanying efficiencies and economies of scale are often available with larger enterprises. But one of our goals with *Start Your Farm* is to responsibly inspire as many *new* farmers as possible. We are convinced that more new farmers will find success with smaller-scale operations than with large, and that's why we've tailored the information in this direction.

Next, a word about production methodology. Though we strongly emphasize holistic and sustainable systems, we understand that ideas about "best practices" vary from crop to crop and from region to region and can also be highly influenced by demographics and distance to markets. Pragmatically, what works for a farmer who is two hours outside of New York City will be completely different (economically, ecologically, and culturally) from what works for a farmer who is on the western edge of Nebraska. These wide-ranging differences present important challenges for us to address. Today, as we deal with problems ranging from climate change to food accessibility to nutrition-related health issues to a growing global population, we would go so far as to say that this gulf between farming methodologies may be one of the great looming issues of the twenty-first century.

As challenging as it is to provide useful, translatable advice for so many different types of farms, we firmly believe that it's in the best interest of *all* producers to strive toward both environmental and economic sustainability and to live within their own means and capabilities. This overarching philosophy certainly influences the structure of the book. Because of this, we find it natural to compare and contrast our methods of farming with examples that we feel are less suited for small and midsize sustainable operations and to clearly explain why. This isn't intended to censure methods that are different than ours, or espouse "small" over "big" in any way. Rather, it's to show that legitimate alternatives to the infamous twentieth-century farming mantra "Get big or get out" now exist. In short, we believe in a landscape of agricultural abundance, not scarcity—and this includes an abundance of ideas.

Sustainability is the common thread. But beyond any one particular methodology, *Start Your Farm* distills practical rules and time-tested wisdom for consistent, successful results. We aim to steer new farmers onto a wide agricultural highway, one with many "on" and "off" ramps—not just one or two. But regardless of how you choose to farm and what you choose to grow, it's our hope that you will be left with a strong sense of what unites us as farmers rather than what divides us.

Ultimately, our goals for this book are simple: to encourage you to think critically, to become an expert observer, and to challenge yourself to keep one eye fixed on profitability. These are the traits of the finest farmers we know, irrespective of what they grow or where they live. By the time you reach the final chapter, you will have explored so many questions and ideas that any pre-conceived notions of what a farm should be or how a farmer should act likely will have shifted. For the novice dreaming of starting a farm, this is probably the greatest service that we, as authors, can provide.

Now it's all up to you. We hope that this book will inspire and embolden you to start the farming journey yourself, and see where it leads. It's a worth-while life, one filled with endless hours of hard work but also with boundless joy and satisfaction. Once you begin, growing food for a living will positively change you forever.

. CHAPTER ONE .

WHY BE A FARMER?

Forrest

If you've picked up this book, you've probably already dreamed of becoming a farmer, growing food not just for yourself, but for your greater community. You yearn to work with the soil, feel the earth in your hands, and are ready to accept a life of daily chores, intellectual challenges, and uncertain financial returns. You're not intimidated by the possibility of seven-day workweeks, are physically capable of hefting forty-pound hay bales or irrigation pipes, and have the emotional resilience to endure an entire year—or five, as can happen—before you make your first profit. All that's left is to trade in your suit and tie for sturdy boots and a dilapidated hat.

Still interested? Allow us to be the first to say "thank you." The world needs your help. Moreover, congratulations! Managing a successful farm is a deeply satisfying, joyful experience, and taking this first step—through self-education—is key.

Farming is one of the most ancient and important jobs on the planet. It's no hyperbole to say that advancements in agriculture have almost single-handedly carried our civilization forward from the Stone Age, ushering in a modern age of food abundance and, consequently, personal free time that the world has never seen before. To wit, as recently as one hundred years ago, nearly 40 percent of Americans were full-time farmers. That's right, 40 percent! Today, it's less than 2 percent. Instead of being farmers, 98.4 percent of Americans now have the freedom, or luxury, to pursue another career entirely, confident that their next meal will be available whenever they choose. This is truly an unprecedented period in human history.

Over the course of the twentieth century, nutritious, reliable, and abundant food transformed our entire globe, sending population rates booming skyward, extending the average human lifespan, greatly reducing disease and famine, and even sustaining peace. While at first glance, one might think these advancements have occurred, respectively, through medicine, technology, or military influence, the actual reason lies primarily with food security. In order to do their jobs, scientists, doctors, inventors, and soldiers alike must eat.

My own grandfather, a career farmer born in 1903, was granted deferments from both world wars because he was considered too valuable at home, growing food for GIs and civilians alike. This is an ancient tradition. Two thousand years ago, the Roman poet Virgil was tasked with writing *The Georgics*, a poem to persuade warring soldiers to return home to become farmers. Even George Washington, retiring to his farm after leading the American Revolution, wrote, "I had rather be on my farm than be emperor of the world."

Yet, as fundamentally important as food is—an absolute necessity for life—there's no guarantee whatsoever that people will choose to become farmers. Strange, isn't it? When I was in college, guidance counselors encouraged students to go to law school or pursue an MBA to start a high-tech company. On career day, recruiters from Coca-Cola, General Electric, and Pfizer flooded the campus, pitching entry-level jobs to enthusiastic marketing, business, and economics majors. Each year, a famous accounting and consulting company hired a small army of my classmates, freshly graduated twenty-two-year-olds, and trained them to give business advice to Fortune 500 companies.

Already intent on being a farmer, I watched, perplexed, as several of my friends received fifty-thousand-dollar salaries straight out of college. My own knowledge of how to make money revolved around harvests of apples or pickup truckloads of sweet corn. When I heard about my friends' salaries, I was genuinely curious. "How can a new graduate get paid that much?" I asked. "And where does that money even come from?" A few years later, my question was answered. That popular accounting company, the one that hired dozens of my classmates, imploded. It was the auditor for a large company called Enron and was caught up in one of the greatest corruption scandals in United States history.

Of course, this was only one example of a mainstream occupation, but it left a huge impression on me and got me thinking about bigger issues—questions of economic stability and the role of corporate and personal ethics. That said, enthusiastic as I was to start farming, I wasn't naive enough to believe that companies would show up at my college to recruit young farmers, or that such businesses even existed.

The takeaway was loud and clear: No one on the corporate end prioritized the professional training of young farmers. And, from what I could tell, back in the early 1990s at least, farming wasn't even a consideration for most recent college grads. But I continually went round and round with one question: *With everyone needing to eat, who is going to feed us?* For an entire generation—an entire culture—to turn its back on farming, I rationalized, there must have been some key element I had overlooked.

Like any budding scientist (and a good farmer is undoubtedly a biologist, ecologist, and geologist wrapped into one), I needed to test this hypothesis. What better laboratory than a farm community? Returning home to the Shenandoah Valley, I spoke with my friends who had graduated from larger universities, programs that, unlike my liberal arts college, offered degrees specifically in agriculture. They told me how their classes had focused on large-scale, chemical- and machinery-reliant operations geared toward industrial-scale farming: corn, soybeans, wheat, and confinement animal production. When I asked about small to midsize farms, or sustainable and organic agriculture curriculums, they shook their heads. By and large, their professors were dismissive of alternative models for agricultural success. They were taught that economics was the most important consideration, and when it came to scale, big was better—and bigger was best.

Next, I spent lots of time with local full-time farmers, men and women who operated the dairies, ranches, orchards, and grain farms in my local community. Asking for advice, I was flat-out informed by several farmers that making a profit in farming—"especially these days"—was next to impossible. In fact, they said that a college degree should be a passport *away* from agriculture and that I'd be crazy not to pursue a different career while I had the chance. Even my own father, a lifelong government employee who worked in Washington, DC, discouraged me. "How will you make money? Pay for health care? Take a day off?" Despite my enthusiasm for farming, I had to admit that I couldn't answer these tough questions.

Looking beyond my agricultural community to society at large, perceptions about farming and food were clearly at odds. Back then, on television, farmers were portrayed as overalled men in distant Midwestern locales, uneducated folks with country accents driving beat-up pickups. They performed vague, humorous-sounding tasks like "slopping the hogs." The relationship between farmers and their crops seemed fully disconnected, as though food simply originated at the grocery store or from a bag at the drive-through. When I brought up farming out in mainstream society, I discovered that most conversations began and ended with the price or the taste of food, rarely digging into where it actually came from.

I struggled with this cognitive dissonance for many years. At turns, I was frustrated and disillusioned, even verging on cynicism. And often for good reason. In the early days of my career, heading home from a farmers' market with a paltry fifty dollars, it was hard to stomach seeing the lines of cars at McDonalds or the vehicles overflowing from the nearby supermarket parking lot. But, as the comedian-philosopher George Carlin once said, "Scratch a cynic, and you'll find a disappointed idealist underneath." In my heart, I knew that I was an idealist, but a practical idealist. I was on a mission to save my family's farm, and I wasn't going to let cynicism—or anything else—stand in my way. Yet, for the better part of a decade, I had little idea of how to actually *do* it.

It took many years, but I finally had an epiphany. The enormous responsibility of feeding our society, so that each and every one of us can live our lives as we currently do, fully hinges on a legion of independent, altruistic *volunteers*. Defying conventional economics, purely on their own volition, farmers willingly devote their physical energy, money, and especially their land—with no guarantee whatsoever of financial compensation—to produce food for their fellow humans.

In a way, it was beautiful to know that such people existed. But it was also shocking when I realized how fragile the entire system truly was. I remember being left with so many questions. What compelled them? What drove and motivated them? Surely, it wasn't money or job security—I had known far too many cash-strapped farmers over the years. Yet, as a society, we were completely reliant on a group of hard-working people who raised their hands each year and said, "I'll do it. I'll volunteer for this job."

It was a mystery that I couldn't begin to fathom. But it was perhaps that very mystery that made me want to join in as well. I understood that no one was going to hire me, salary me, or give me a raise. If I was going to farm, I'd simply have to be stupid or stubborn enough to commit to the job.

If it seems strange to think of farmers as volunteers, then think about these questions: In what other occupation do people undertake a job with endless hours, no guaranteed pay or days off, and no realistic likelihood of retiring? And what other job is so fundamentally important for society but remains so consistently misunderstood? The more we dig into these questions, the clearer it becomes that one would have to be intentionally defiant, even borderline contrarian, to sign up for this career. Running a farm is an endurance race that has no finish line.

Farmers, it turns out, are a very special subset of the population. The sacrifice involved in making this commitment is foundational; it means willfully turning one's back on the mainstream. As Wendell Berry eloquently states in his poem, "The Contrariness of the Mad Farmer,"

[. . . .] If contrariness is my
inheritance and destiny, so be it. If it is my mission
to go in at exits and come out at entrances, so be it.
I have planted by the stars in defiance of the experts,
and tilled somewhat by incantation and by singing,
and reaped, as I knew, by luck and Heaven's favor,
in spite of the best advice.

Being a farmer, as Berry makes clear, is more than just a job description. It's conscientiously choosing a different way of life—a willful, intentional defiance of cultural norms and expectations. Once this journey is begun, there's typically no going back. But that's okay. After your first taste of success, you won't want to anyway.

Still, in the face of these daunting cultural and economic challenges, it begs the question: Why be a farmer? The answer, to be sure, is complicated. I could spend page after page praising the merits of fresh air and sunshine, of rich soil and pure spring water, of honest hard work and abundant harvests—and each would surely be a worthy reason to devote yourself to a life of agriculture. But this wouldn't do justice to hard realities such as paying for land, supporting a family, planning a budget, securing health care, or any of the myriad unromantic daily details that contribute to farming success.

Alternately, I could tell you exactly why I chose to farm, placing my degrees on a now cobwebbed shelf, setting off to rebuild miles of dilapidated locust fence, digging ditches for water pipes beneath a towering July sun, and returning home each night exhausted, penniless, and smelling of pig manure—yet sublimely satisfied. At the end of the day, though, this remains my story alone, circumscribed by my own idiosyncrasies. Certainly, there are lessons to be shared here, and I'll share them. But it's impossible to farm exactly like someone else. One-size-fits-all farming has always been a recipe for eventual failure.

Instead, there are far better reasons to be a farmer—reasons that specifically involve *you*. Farm because you hear the call, and you must listen. Farm because a nagging voice inside your head insists, *You have to do this. You have to try.* Farm because, despite the mountain of evidence to the contrary, you are convinced that you can succeed. In your heart, you *know* you can do it.

Understanding why you should be a farmer will forever be a personal process, one that will unfold as you apply the information contained here specifically to your unique circumstances. No one but you can fully explain it. But when the day arrives when you've grown your own food, earned your own paycheck, and glimpsed what it means to be self-reliant, perhaps words will no longer be necessary. Certain sweet moments in life transcend description.

THE LANGUAGE OF SUSTAINABLE AGRICULTURE

From the get-go, we want to share some strong opinions that we have about sustainable farming, lessons we've cultivated over many decades. We'll use the rest of this chapter to define and espouse important terms and concepts and to spell out how they apply to contemporary agriculture. Then, in the ensuing chapters, we'll investigate these topics more thoroughly. As you pursue your own farming dreams, keep these themes in the front of your mind. Following this advice might not guarantee success, but it's certain to put you on a firmer path to economic and agricultural sustainability.

Let's be perfectly clear: The cultural, social, and economic headwinds you will face as a farmer are strong and sustained, and they aren't likely to diminish anytime soon. To ignore these realities is to do so at your own peril. Illustrating this point, take a look at these two graphics:

US Farm Jobs as a Share of Total Jobs, 1790-2014

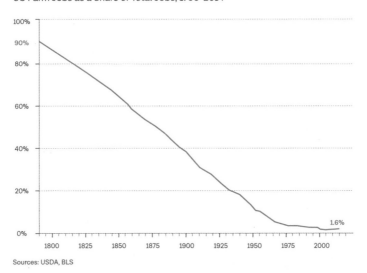

Sources: USDA, BLS

Farms, land in farms, and average acres per farm, 1850-2012
Million farms/billion acres/hundred acres

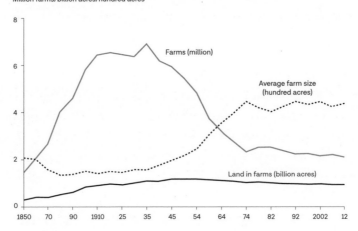

Source: USDA, Economic Research service using data from USDA, National Agricultural Statistics Service, Census of Agriculture.

The first chart shows a steep historic decline in the percentage of the working population that farms for a living, and the second, a steady reduction in available farmland. This certainly seems inauspicious for a burgeoning career in agriculture. Worse yet, it's easy enough to extrapolate corollaries: higher real estate prices, additional mechanization, and a greater need for capitalization. Combine all these factors, and it's easy to see why starting a farm today can seem next to impossible.

That said, flying in the face of these daunting statistics, we believe that these very facts are actually wonderful reasons to *start* a farm—and, furthermore, that the timing to start a career in agriculture—right now—probably couldn't be better.

Let's review that first chart with fresh eyes. True, we presently have fewer farmers than ever before. But overall, the United States has a growing national population, we all require a daily balanced diet, and over the past several decades, the value of fresh, nutritious food has become widely accepted. See the opportunity yet?

Now, study the second infographic. True again, farmland is diminishing. But because of encroaching housing development, consumers, towns, and cities are now more accessible to farmers than they've ever been—or are about to be, depending on where you happen to live.

Time for an easy equation: a large, local customer base (demand) + a decreasing number of experienced food producers (supply) = a dream scenario for astute, entrepreneurial farmers. It's no accident that some of the world's most successful investors presently view farmland as one of the best investments over the next several decades.

The point? Just because something seems intimidating or fraught with difficulty is not a reason to dismiss it out of hand. In fact, the opposite can often be true, which leads to one of the fundamental tenets of this book: *Always view problems through a lens of opportunity.* With a little tweaking and perspective, even our biggest liabilities can be coaxed into assets. As sustainable farmers, this mindset is imperative. When it comes to farming, a lifetime of daily challenges is guaranteed. To counterbalance this, we must identify and carefully consider every reasonable opportunity that comes our way.

Next, let's define a word we've already employed several times: *sustainable*. Like many terms in farming—e.g., organic, biological, commercial, local, industrial, heirloom—this word can sound overused and ultimately generic, a catch-all term that makes for a nice sound bite but otherwise doesn't explain much. After all, wouldn't every farm want to be sustainable? (Indeed, we'd argue that all farms, regardless of scale or enterprise, should strive toward this goal.) Yet with so many different types of farms, and with operations spanning all corners of the continent, why should this specific word remain especially relevant?

Typically, the concept of sustainability is associated with environmentalism; for example, one might say, "a truly sustainable business creates a net-zero environmental impact on the local ecosystem." Much of the book will focus on environmentally friendly and ecologically focused solutions, with the creed *maximize what's free* (most notably, sunshine, rain, soil carbon, and airborne nitrogen) being a recurring theme. But while environmentalism certainly plays a major role in overall sustainability, it's best to visualize the word itself as a three-legged stool: Take away any one of the legs, and the stool tips over. So while *environmental* considerations are certainly one leg of the farming sustainability stool, there are two more *E*'s that make up the other legs: energy and economics.

BIOLOGICAL FARMING VERSUS ORGANIC FARMING

Organic farmers are certified by the USDA; which fertilizers and other agricultural products they can or cannot use is defined by law. A biological grower may not be certified organic, but follows the principle of putting healthy soil biology first, and not using synthetic chemical inputs—in essence, following organic protocols without the formal certification. While the terms *organic* and *biological* are often used interchangeably, the way we use the word *biological* in this book is more closely synonymous with *sustainable*.

By *energy*, we mean human energy, and more specifically, your personal energy. It's critical to have an honest understanding of your physical and emotional energy levels, living within your capabilities and not overextending yourself. Burnout is high in farming. This doesn't mean that you shouldn't push yourself or challenge your preconceived notions about what you can accomplish. But farming is typically a seven-day-a-week job, year-round, and this is especially so for the new farmer who is unable to afford hired help. Keeping your own energetic sustainability in mind—this means taking breaks, eating nourishing food, getting a reliable night's sleep—will be crucial to prolonged success.

The third *E* is *economics*. This is often the toughest element of sustainability for new farmers to understand, and we'll spend no shortage of ink exploring this topic from multiple angles. Often, the temptation is to simply focus on the first two *E*'s, environmentalism and personal energy, hoping that the economics will mysteriously take care of themselves. It's like that old adage, "Do what you love, and the money will follow." While this might occasionally work, when it comes to farm economics, however, a more reliable aphorism is this: "Fail to prepare, and prepare to fail."

Put a slightly different way: *Rely on no one but yourself for the economic success of your farm.* You can't expect anyone else to have the passion and dedication that you have for your land. This isn't a job you can outsource. You must be the one who figures out the dollars-and-cents equations of your operation, treating it as any other business person would. After all, if you don't pay your bills, you don't get to be a farmer. Period. As the personal expert of your finances, you become one of the most sustainable assets you'll ever possess.

This leads to another point that we will repeatedly highlight: *profitability.* Simply stated, being profitable in farming is selling a good or service, then having money left over after all expenses, including paid time, have been deducted. (There's one major caveat to this, which we'll address in chapter 5—the difficulty of recapitalizing land expenses.) Regardless of where you live, what you grow, or which agricultural methods you use, if you don't make a profit, you won't stay in business for long. As alluded to earlier, running at a loss for the first five years or so is actually a reasonable expectation, and you should take measures to prepare for that. But before long, profitability (and profitability at a reasonable rate of return—again, we'll discuss what this means) must occur. Farmers must make every attempt to avoid the purgatory of always breaking even, or worse, perpetually operating in debt. Which brings us to the next point: *debt.*

Avoiding debt should be a primary goal for any new farmer, even if you have to start very, very small for a few years. This is how I operated my farm for close to a decade, only spending money that I had made through sales of farm products, and not borrowing extra. This pay-as-you-go strategy eventually allowed me to purchase forty-two more acres, all made possible by a dozen (many, *many* individual dozens of) eggs at a time! But over the past fifty years, debt has tanked more farms than drought, flooding, and pestilence combined. If there's one lesson we've learned from our endless series of national economic bubbles (housing, gold, oil, stock market, savings and loans, to name a few), it's how financially debilitating debt can be for the average person. Farmers aren't immune to these challenges. Legions of great producers have been forced to relinquish their farming dreams simply because they couldn't pay their debt when the bank came calling.

Borrowing money (almost always with interest) allows you to hyperaccelerate your goals, turning dreams of tomorrow into realities of today, but with accompanying risk: extra years of paying back compounded interest, or worse yet, losing your collateral (often the land itself) if you're unable to pay back the loan. While borrowed money might buy a tractor, a new barn, or even the land you'll be farming, *experience*—the most valuable farming asset of all—cannot be purchased.

Experience doesn't come with a bachelor's degree in agriculture, and we'll be the first to insist that it doesn't come from a book. There are no shortcuts; it must be slowly acquired over time. As you've already discerned, agriculture is replete with uncertainties, surprises, and intellectual challenges—and that's just before lunch. Adding monthly debt payments to this intimidating list financially handcuffs most producers right from the start. Combine that with lack of experience, and you've greatly increased your likelihood of economic failure.

So does this mean that you should never take on debt? Certainly not. There are plenty of times when leveraging assets makes sense, especially later in your career. As you gain farming experience and create reliable cash flow (the steady inflow of revenue into your bank account, meeting or exceeding your outflow), these opportunities will become clearer. We'll discuss them more in later chapters. In the meantime, however, take this generalization to heart: Avoid debt as much as possible.

Another theme we'll focus on is *failure*. Wait, not succeeding is a topic of discussion in a book about how to succeed? Ironic, I know. Our culture is utterly obsessed with the topic, simultaneously terrified and captivated by it. Personally, I know a handful of folks who spend their days avoiding the so-called humiliation of failure at all costs. Some of these people fear failure so much that they never try to accomplish anything. The thought of making any kind of mistake, no matter how small, paralyzes them.

If failure is a major concern to you, here's a spoiler: In farming, you *will* fail— 100 percent chance. In fact, with apologies to Benjamin Franklin, failure on a farm is every bit as certain as death, taxes, and Red Sox fans hating the Yankees.

But here's what no one ever told me: It's okay to fail. Moreover, in farming, it's important to fail. While painful at first, failure can be an enormously useful tool. It helps us learn our personal limits of time and energy. Making mistakes is an instrumental timesaver in the long run, clarifying what works well and what's a complete boondoggle. Failure provides us perspective and invaluable experience (there's that word again) for future enterprises, making us intellectually stronger and more emotionally resilient.

So thumb your nose at that sagging bookshelf loaded with self-help books that insist you're not a failure. Yes, you are! Get out there and make a bunch of mistakes. But while you're doing so, fail properly. Fail gracefully and thoughtfully. This might seem like yet another contradiction, but it's the only way to recognize success when it finally arrives.

Now here's a related tip: *Don't worry about what other people think.* I began this chapter editorializing about the cultural disconnect that we, as farmers, often face with the public at large, and frequently even with our peers. This point was repeatedly driven home for me early in my career and helped shape my thinking as a new farmer.

In 1994, when I was twenty years old, I found myself talking to an older farming couple at a local picnic. Like me, they raised cattle for a living, but like most commodity cattle farmers, they sold their animals straight to the livestock auction, bound for corn-fed confinement feedlots. When they asked me about my farming ambitions, I told them of my dream to sell 100 percent grass-finished beef. I explained how the cattle would be completely organic and that I'd directly market the meat myself to local customers. I told them that my farm could provide food for several hundred families once I really got going, and perhaps we'd make a profit for the first time in a generation.

Their reaction? When I had finished speaking, they turned to each other, made eye contact, and burst into uncontrollable laughter.

Decades later, despite this withering response from my elders (they apologized for their behavior after they managed to stop laughing, bless their hearts), my farm has accomplished all of these goals and much, much more. If I had worried what my neighboring farmers thought of me, I certainly wouldn't be sitting here now, writing this book. As Eleanor Roosevelt once said, "You wouldn't worry so much about what others think of you if you realized how seldom they do." In short, believe in yourself, and just go for it.

As for that couple? A few years ago, they put up a sign at the end of their lane: "Free-Range Beef for Sale." Pardon me while I indulge in a moment of laughter.

All that said, even if you become a superior producer, your products won't sell themselves. That's why we'll repeatedly focus on heading to market in chapters 7 and 8. Perhaps you want to keep honeybees or grow cantaloupes or start a sauerkraut business. Terrific. Maybe you just want to sell wool to local knitters. Again, wonderful. I like honey, sauerkraut, and knit caps as much as the next guy. But how are you going to find customers? Do they live in your neighborhood, or a thousand miles away? How much will they buy? When you expand your business, how will you find more customers? What will you do if they buy *all* of your products and you're sold out? What will you do if they buy *none* of them and you have an entire warehouse full to sell?

Before you start that first hive, jar your first kraut, or shear your first ewe— and certainly before you plant hundreds of acres of corn, like we used to do on our farm—you'll need to invest time (lots and lots of time) figuring out where you're going to sell your products and how you're going to get them to market. Then, once you've done this, you'll need to create a backup plan. Even then, just to be safe, you should come up with yet another backup plan. Chances are, you're going to need both of them.

Whether you're a niche farmer growing a specialty product or running a large-scale operation, smart producers spend an enormous amount of effort finding and retaining their customers. This is every bit as important as growing the food to begin with, because without appropriate sales channels, cash flow will dry up, fresh produce will spoil, and customers will find products elsewhere. When all those cantaloupes ripen at the exact same moment, you'll need a place to sell them, and fast. Have a solid marketing plan prepared well in advance.

Your location, of course, will have a strong impact on how you're able to best market your products. Still, no matter where you begin your farm, or what you decide to grow, it's vital to *match the land to its suited use*. Rainfall, soil types, topography, and climate zones will all have a tremendous impact on your success. Here, we should take our cues from nature, even utilizing bio-mimicry. On my farm, for example, wild turkeys, deer, cottontail rabbits, and raccoons naturally flourish. As such, it's no coincidence that I'm able to raise free-range chickens, sheep, cattle, and pigs on the land. While the correlations aren't identical, when I stand back for a moment and consider the landscape, there's a nice pattern here—omnivores and grazers, birds and mammals, animals big and small. As farmers, we can try to force our human visions onto the land, or we can have the wisdom to work with what nature gives us.

Finally, *have a sense of humor*. Lighten up, Francis. When it comes to farming, it's only a matter of life and death.

Take an average day at a mainstream job. What's the worst that typically happens? A client becomes irritated or the supervisor reams out a coworker. Maybe Larry (whatever happened to guys named Larry, anyway?) gets his tie caught in the document shredder . . . again. Hey, somebody get that guy a golf shirt!

But on any given day on a farm, things *die*. And not in any noble, dignified, or discreet kind of way, either. Things die screaming, eviscerated, and—more often than we'd care to think about—partially masticated. Have you ever walked through the morning dew to check on your free-range chickens (cue love theme from *St. Elmo's Fire*), crested a hill, and found them slaughtered willy-nilly (cue Metallica), their gleaming entrails spilled across the clover?

Frankly, such eventualities put this whole farming thing into perspective. And faced with the possibility of daily mayhem, pestilence, and literal plagues of locusts, a sense of humor can be a handy-dandy coping mechanism.

I learned this particular bit of wisdom from Travis, a farmhand of more than fifty years. Travis arrives on my farm each morning sporting a nonironic trucker's hat, unruly lamb-chop sideburns, and an emotional disconnect that leaves no doubt that he's capable of neck-punching someone into a coma. After frantically stomping out a brush fire one afternoon, we stood with hands on hips, breathing heavily, melted boot soles still smoldering. Somehow, we had snuffed the flames scant feet before they had ignited our neighbor's tinder-dry cornfield.

"You know," he remarked after a moment, "if we didn't laugh about these sorts of things, we'd probably just end up murdering each other."

Right you are, Travis. Right you are.

All of which leads us back to the main question of this chapter: *Why be a farmer?* As you've probably figured out by now, the answers are far more complicated than they might seem at first glance. They're rooted as much in psychology, sociology, and economics as they are in science, soil, and sunshine. Farming, after all, requires intellect as much as brawn. As we move forward, the answers will sometimes appear contradictory, sometimes counterintuitive. Farming is for neither the faint of heart nor the intellectually lazy. For you—you, personally—the true answers to why be a farmer might only be revealed long after the final page of this book is closed.

By now you might be thinking, *Hey, I haven't read a single paragraph about tomatoes, tractors, or goats. Where's the talk of sparkling spring mornings, beautiful rural vistas, and glorious evening sunsets? What kind of farming book have I gotten myself into?* If something along these lines is nagging you in the back of your mind, then that's understandable. We'll get there.

But if you've made it this far, then you've already taken your first important step: redefining your expectations of owning a farm business. In the twenty-first century, this means casting aside 99 percent of what society tells us farming should be and learning about what it actually *is*.

Still want to be a farmer? Congratulations! You're entering a world of excellent company. As Bob Evans—yes, *that* Bob Evans—once told me, there's no finer group of people on the planet than those who call themselves farmers. It's a life of never-ending challenges and hard work, but also one of incredible joy and satisfaction. By all means, join us.

CHAPTER REVIEW QUESTIONS

1. Are there farmers in your area? If so, what type of agriculture do they practice? If there aren't, why do you think this is the case?

2. Do you know any farmers personally? If so, what are their opinions on the future of farming? If you don't, how do you plan to meet and interact with full-time farmers?

3. What is your farming dream? Describe it in detail. Explain why *you* are the person best suited to manifest this dream.

. CHAPTER TWO .

DO YOU HAVE WHAT IT TAKES?

Ellen

One of the things about farming that makes it endlessly interesting—and almost never mastered—is that it involves so many skills. A farm requires intense physical work, mostly performed outside, in all weather conditions. It's also a business with a real-world financial stake and multiple relationships to juggle. At the same time, it's a tangible, hands-on place, with buildings, wells, wires, posts, and machines that all need to function properly. In this chapter, we will take stock of what you have in your corner, and what you may either need to acquire, learn, or buy.

You are embarking on a complex and widely encompassing journey, and this requires an honest assessment of what your strengths and weaknesses are as you approach your farm business. This chapter will be a guided tour of various arenas of self-assessment, an exercise that should spark quite a bit of internal dialog, and possibly some exciting conversations between you and your potential partner(s)—both business and life partners. Certainly, there are professionals of all stripes that can be hired to perform almost every function on a farm. But you will need to handle the majority of them yourself, in-house, or there just won't be enough core knowledge to hold your business together from one day to the next.

SKILLS ASSESSMENT

Business Experience

Have you worked in a successful business before? This question appears straightforward at first, but, upon deeper inspection, it begs the next question: What defines success? In the broadest sense, a successful business has financial stability, and it lasts as long as its owner wants it to endure. But, just as important, it brings joy and satisfaction to everyone it touches: owners, customers, employees, suppliers, and service providers. As you conceive of your farm business, consider what success really looks like to you. If you have not experienced working in or for a successful business, imagine what that might look like. If you have, try to recollect, in detail, what made that business a great one.

If you've already run a successful business, wonderful. This experience is a tremendous asset that you can capitalize on immediately. Even if you have run an *unsuccessful* business in the past, then you have already learned lessons that will help you on this next adventure. If you haven't run a business at all, then buckle up and keep reading.

Do you have any experience with accounting? Have you kept the books for your personal accounts or for a business? Any work experience relevant to balancing weekly finances (and yes, this means cross-referencing each bank statement as it arrives) will help when you set up your farm business. If the idea of numbers at all disturbs you, then this is going to be a painful uphill slog. Not only must farmers carve out the time to sit down and tally the numbers but they must also tally those numbers correctly. I jumped headlong into book-keeping and found that I actually enjoyed it. I think you might, too. Having a positive, can-do attitude is the first important step!

Production Capabilities

You're about to start growing things for a living. This means, by definition, that you'll have to have a green thumb and perhaps a way with animals as well. At its core, a farm harnesses the power of the sun, transforming it into food, feed, flowers, and fiber (e.g., wool, cotton, hemp). You will you have to be good at not only growing plants but also managing Mother Nature's free resources: sunlight, rain, soil, and carbon. In other words, your skill set for "growing"— regardless of what you ultimately produce—must be high.

Think about what you've grown well so far, and how you know that you did a great job. You'll be competing in the marketplace with all kinds of other growers, from old-time backyard gardeners to full-time professional farmers. If you're going to make it, you need to know what the standards are—what "good" is. This means witnessing with your own eyes what an excellent crop of tomatoes or a healthy brood cow or high-quality hay looks like. These are things you must know, and, to do so, it's crucial to get yourself in the presence of excellent examples. Before you plant your first seed for commercial purposes, you should travel to other farms, gardens, field days, seed trials, and open houses. Get up front, ask questions, and begin to collect and fathom some answers.

For most budding farmers, it's the growing part that gets them thinking about having a farm to begin with. If you've gotten this far, you probably already like working with plants or animals. You like being outside, being active, getting dirty, and watching things grow. To turn this into a successful business, however, you have to be good at this—really good. If you don't have lots of high-quality product coming off of your land, even the best systems will soon fall apart. It won't matter how lovely the homestead looks or how beautifully you can set up a farmers' market display or how many wonderful spreadsheets you can create to keep track of your sales and expenses.

This is why we strongly recommend working with an experienced farmer early on. Knowing how to grow things well is the kind of skill you want to get paid to learn, because doing it on your own dime can quickly become very expensive! When you work for a grower, you get to watch them make mistakes and observe their successes. You can then take that experience and knowledge with you into your own farm business. The good news is that such training opportunities abound, and we will outline them for you in chapters 4 and 5.

Marketing Capabilities
Selling what you grow creates income, plain and simple. If the words "selling" or "marketing" give you the chills, take note. These are verbs—more specifically, *action* verbs—and they require hard work and skill. What experience do you have, and can you get excited about becoming a seller/marketer?

Needless to say, as a new farmer, you will be wearing both hats—lead salesperson *and* head of marketing—at least at the beginning. In myriad ways, you will represent your farm to the outside world. You'll want to have your "elevator speech" polished and ready, so you can quickly and powerfully deliver your mission statement to anyone who might ask about it. Passion is persuasive, and authenticity is your ally. One of the most trying places to keep your cool and maintain eloquence is when you are face-to-face with a customer who is questioning your prices. Make no mistake: This is a PhD-level communication workout, where you must calmly but confidently explain how you've arrived at your price, the value of your product, and why someone like them would really benefit from buying it. If you can manage to pull this off, you might just find yourself with a loyal customer for life.

In this age of social media, telling your farm story with photographs and engaging writing has become pretty much mandatory. Your new farm will need online promotion, which will work best coming from you or from some-one very close to the farm. It's difficult and expensive to outsource this one. And customers today have a good eye for authenticity.

Do you have direct experience with customer service, or have you paused to consider what great customer service means to you? A prior role in a suc-cessful business will have given you lots of access to this knowledge. But if you haven't worked in sales or marketing, we suggest that you get some training right away. Don't defer this important component of your business education! One of the best ways to learn is to find a grower who will let you shadow them at their various market outlets—at market, on a sales run to a restaurant, or as they negotiate a wholesale contract with a prospective buyer. Of course, you will need to explain your goals to the grower, and if they are confident in their own marketing prowess, chances are good that they will be generous with their knowledge. Always offer to pay them for their time, whether with money or labor. It's the courteous thing to do, and you're going to benefit financially from the experience.

PERSONAL TEMPERAMENT

There are as many ways to farm as there are farmers. While there are a few absolute musts—and the occasional never-nevers—most of the time, multiple paths exist to achieve the same goal. Any given farm, then, becomes a direct reflection of the skill and temperament of the farmer. It's vital to have a clear understanding of your character traits and an honest appraisal of your per-sonal nature. You can always bring in other personality types to your farm and hire individuals with characteristics that you are lacking. But, in the end, it's *you* who will set the tone for your farm. The importance of your personal temperament will have wide-ranging ramifications over the course of your career. Let's take a look at a few key traits that will influence the nature of your farm.

Leadership

A farm is a complicated creature, comprised of multiple moving parts—all kinds of living beings and lots of systems to get things done. In many ways, your farm will be like a ship alone on the ocean; for steady, smooth sailing, you must be its capable captain. My experience on my own farm—as well as my experience consulting for farmer clients—is that this leadership component can be the hardest to master. It's also a skill set that farmers rarely spend time or devote attention to learning, especially at the beginning of their farming lives. But by about year five or so, once you've begun to hire more help, a lack of good leadership will become one of the biggest limiting factors to your farm's success.

A good leader is one who knows the business from all angles and can set priorities, and then effectively communicate them to the team. Good leaders influence others by setting an example, teaching them how to think about themselves, the work, and the larger community. Finding and developing the aptitudes of each team member, by striking the balance between challenging and inspiring them to do their best work, is the ultimate goal of the leader.

Traits of capable leaders include emotional intelligence, flexibility, vision, and the ability to engage others in that vision. Are these traits that you currently embody, or are they traits you can learn? I

SELF-DOUBT

I know a farmer who continually has trouble keeping employees happy over the course of a whole season. While he has vast experience in his enterprise, he second-guesses his own daily and hourly decisions in front of his team. Naturally, the team begins to lose confidence in him as a knowledgeable leader. He also has a habit of changing the work plan multiple times each day. Folks get confused about what to do, how to do it, and when to do it. There is an underlying sense of chaos and uncertainty. This is a recipe for unhappy and disrespectful employees, resulting in poor team spirit and low productivity.

believe that, in this area, trying counts. Having the intention of becoming a good leader will keep you in good stead as you gather the skills and tools to actually become one.

Decisiveness, Flexibility, and Planfulness

Farming is an ongoing process of making long-range plans, from which come detailed action lists. A daily or hourly plan is made, then perhaps the weather changes or an animal gets sick or a tractor won't start. Immediately, plan B needs to come to the fore. This sort of thing happens on a regular basis, and the farmer needs to balance the commitment to a well-conceived plan with the ability to change and adapt at a moment's notice. In other words, you can plan on needing a lot of plan Bs!

This is what we refer to as planfulness. You have to keep rolling with ever-changing data. A farmer with good foresight and planning skills will inspire others and create an atmosphere of order and low stress. In general, people, plants, and animals abhor chaos. It's incumbent upon the farmer to reduce disorder as much as possible. Remaining the level-headed captain promotes calm, fosters morale, and establishes an important precedent of "how we do things around here." Even if you never need it, always keep a plan B handy in your back pocket.

But you will have to balance planful flexibility with decisiveness. If you have ever built a house, or even simply redone a kitchen, you've had the pleasure (or arguably pain) of making an endless number of decisions. The type of tile, the width of a sink, the variety of cabinets, the number of electrical outlets—there are dozens and dozens of considerations. Building a farm is just like that: lots of tiny decisions that mount up over time and come to define your enterprise. If you struggle with making decisions in a timely way, you will be hard-pressed to meet the demands of your fledgling farm while making monthly, weekly, and daily action plans. You have to be decisive, timely, *and* excited about the choices you're making. The decisions will just keep coming.

Joyfulness

Did I mention that farming can be super hard sometimes? Farmers are committed to persevering through heat and cold, rain and snow, wind and humidity, bugs and critters. Throw in some manure, knee-busting stones, and an overlooked basket of rotten onions, and now you have a sense of all the temperatures, smells, injuries, and wetness levels that you will endure. While none of us farmers love all of these conditions, we don't allow them to get us down, either.

We understand that the counterbalance to these trials are the absolute joys that we experience as regularly and consistently as we live, work, and breathe. On a farm, visual treats abound. Every kind of sky and cloud, every color of leaf and flower is represented; bees and worms, beets and lambs are each present over the course of a season. Beautiful sounds, smells, flavors, and textures delight the other senses. These physical pleasures must, over time, outweigh the hardships. A fundamental sense of joy will help tip the balance toward the positive.

Farming is an inherently optimistic act. To plant a seed is hopefulness incarnate! In your worldview, is the glass half full, half empty, or are you pleased as punch that there's simply a glass at all? A joyful temperament will make this journey ultimately more pleasurable and successful. Joy is infectious and inspiring. As the farmer, it is up to you to set this tone on your farm. It will directly affect your teammates, family, customers, and community. Over the course of my career, I've met no shortage of bitter, grumpy farmers along the way. Trust me when I tell you that they will not find success in today's business landscape. People are naturally attracted to smiles, laughter, and positivity. When your farm is a joyful place, the light shines straight through from your fields to your employees and ultimately to your products.

Perfectionism Versus Pragmatism

Many professions call for perfectionism, but on a farm, the perfectionist will meet unending obstacles. Not only are there very few situations on a farm where you can actually get things 100 percent right, but also, just as important, there are few times when getting something 100 percent right actually *matters*. For example, if a few bean plants in each row get hoed out by mistake, the remaining plants will grow larger and fill in the gaps. Thus, hoeing quickly—and accepting a few bean plant casualties—is preferable to going painstakingly slow to ensure that every single bean plant survives. The degree to which you can comfortably find that balance between perfection and "good enough" will largely determine your potential for satisfaction as a farmer. Will you be satisfied with 92 percent correct, or potentially even 84 percent? The profit margin on agricultural products is slim. Therefore, the amount of effort and time it takes to get something exactly right won't, in most cases, be worthwhile. It's just too expensive.

Lucky for us, there is a lot of tolerance built into biological systems. Plants and animals have a spectrum of conditions under which they can survive, grow, and even thrive. We don't have to get everything perfect. Usually, if we focus our efforts on creating a healthy, nourishing environment for our plants and animals, they will take the next productive steps on their own.

That said, limits certainly exist. For example, a tender tomato transplant can't survive freezing temperatures without human intervention. At some point, a plant without enough water will wilt beyond saving. In any farming enterprise, there are extremes that must be remembered, watched, and avoided. But, more often than not, somewhere between pragmatism and perfectionism, most things will be okay. Striking the right balance—and knowing how to pair this to your temperament—is the key.

THIS IS NOT ROCKET SCIENCE

My husband is a mechanical engineer. He designs and builds machines that have tolerances of 0.001 inch or less. He has to get the numbers exactly right or the machine (or project) will fail. His attention to detail and his relentless pursuit of perfection are keys to his success as an engineer. And, he loves that challenge. But put him in the garden, and it's going to take him practically forever to finish any little project—whether it's bed preparation, seeding, weeding, or harvesting. His temperament is not suited to being a professional farmer.

CAPITAL RESOURCES: TIME AND MONEY

Beyond your skill sets and personal temperament are the equally important considerations of tangible assets. You've probably already been thinking hard about what capital resources you can bring to bear toward your farming dream, so let's look at a list of possibilities. While you certainly don't need *all* of these resources to get started and make headway, you will eventually need a good mix of them in order to establish a firm foundation for your farm.

The first two resources are probably the most obvious—time and money. The amount of time you can devote to a farm is next to endless, limited only by how many hours there are in a day; as soon as one project is complete, another half dozen are waiting in the wings. Similarly, your business will indifferently absorb every last nickel you own, and then some. The trick, again, is to find a balance. When it comes to time and money, you have to have either plenty of time or access to money, if not both.

When we say money, we mean cash in the bank—existing funds, and income that may continue to flow in from jobs you (and possibly your partner) hold. You might secure a personal loan from a friend or family member. Your life's cheerleaders may want to help you get the new farm business off the ground and supply much-needed cash at a favorable interest rate. Alternately, you may secure a loan from a commercial lender. Chances are, it will be difficult to qualify for an operating loan—money to pay for your day-to-day operations and supplies—for a brand-new farm. That's because you have no track record to prove your ability to repay the loan. (If you have had the opportunity to begin your farm on an incubator, which we'll discuss in chapters 4 and 5, then you have created a financial history that can help support your claim of being a good credit risk.) If you're in a more comfortable financial position with strong equity in other real estate, you may be able to use those assets as collateral.

Though we strongly encourage you to avoid debt as much as possible early on, by and large, most new growers use their personal savings and family loans to get the ball rolling. Additionally, most beginners continue off-farm jobs, either part-time or seasonal. In my early farming years, I worked a retail job during the winter. I sure didn't love it, but my bank account needed that inflow of cash to float me until the farming season resumed in March. It's never glamorous, but during those early years—when you have more time than money—every penny counts. It's a wonderful year when your farm can finally allow you to quit all those other jobs!

Loans from the federal Farm Services Agency (FSA) are available. These are aimed at helping young and beginning farmers finance the purchase of land and build up infrastructure. But beyond the concern of heaping debt upon your new farm business, to even qualify for these loans, a "beginning" farmer must have between three and ten years of experience prior to application. As we will emphasize throughout this book (especially in chapters 4 and 5), using your time early in your career to grow your skills with an experienced farmer can often be far more valuable than having enough money to buy your farm outright. This push-pull of time versus money will require constant attention and assessment in your early years of farming.

No matter the particulars of your situation, it's incumbent upon you to run the numbers to determine how much money you will need to operate your farm, how much income you will need the farm to generate for your family, or how much you might be able to offset this by providing more of your (probably unpaid, at first) time early on. It is completely normal for farm families in the United States to require off-farm income, as you can see in the chart below. So, as long as everyone is on board with the plan, there is no shame in needing one family member to continue to work "in town" until you become profitable. Many growers obtain health insurance benefits via that off-farm job, which is a sound strategy.

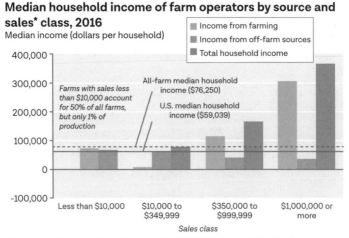

Median household income of farm operators by source and sales* class, 2016
Median income (dollars per household)

- ■ Income from farming
- ■ Income from off-farm sources
- ■ Total household income

*Sales = Annual gross cash farm income before expenses (the sum of the farm's crop and livestock sales, government payments, and other cash farm-related income).
Source: USDA, Economic Research Service and National Agricultural Statistics Service, Agricultural Resource Management Survey, and US Census Bureau, Current Population Reports. Data as of November 29, 2017.

Time is equally as valuable as money, and therefore just as important to gauge. The first question, then, is when do you need the farm to cover production costs, but not your labor? Put another way, how long can you maintain your unpaid farming lifestyle, without the farm contributing to your personal income stream? The more time you have here, the bigger the window you have to develop efficient growing systems on the farm. It takes time to learn how your particular soils work, or how your exact microclimate affects plant and animal growth rates and yields. Learning how to operate tools and equipment to grow crops also requires prolonged practice.

Time can act as a substitute for money, particularly when addressing soil fertility. Should you find yourself on land that is in need of repair and nutrients—which we find is the case on most farms—you can take the time to grow successive green manure crops (a process called "soil building") and let Mother Nature do some of the work for you. With enough time, you can grow your own organic matter (carbon), or you can gradually bring it in from elsewhere in the form of manure, compost, mulch, or hay. Or, if time is more of a constraint, you can purchase all these needed amendments—that is, provided you have the money! (For more on the subject of soil, see chapter 6.)

As a farm consultant, the greatest challenge I face is when an aspiring grower has very little money to invest in operating expenses (e.g., seed, fertilizer, tools), yet needs the farm to supply net income—a paycheck—right away. Talk about a tall order! There's an old joke in farming communities. "How do you get a million dollars from farming?" (Long pause.) "Start with two million." If solving the riddle of money versus time were as easy as many of my clients wished, then no doubt more people would try to be farmers.

Now unless this client has landed onto some very rich soil indeed, soil that will grow plants perfectly without any help, then this is a recipe for disaster. Without sufficient funds to purchase the nutrients the crops will need, you can imagine how well those fields will yield. Without money to obtain fertility, scads of time and effort will go into planting and tending, with very little product to show for it. Unless there's a cash reserve to get you from one season to the next, it's naive to think that depleted soils can provide immediate net income. Many soil scientists suggest it takes five full years to see a significant turnaround in production capabilities in depleted soils. And we think we can sprint to a quick profit by throwing money at it? It can certainly help, but time is still part of the healing equation.

Instead of this next-to-impossible scenario, if you have enough time before you need the land to supply you a paycheck, you can slowly grow the fertility of your land, as well as hone your systems toward efficiency. Conversely, if you have money, you can potentially (and cautiously) buy your way into a success-ful setup—purchasing soil amendments, tools, equipment, and consulting ser-vices. But even this will probably require a year or two in a best-case scenario. Ultimately, both paths can work, and each has its advantages and challenges. The takeaway here is that you must have at least one or the other: You can't have no time *and* no money.

Skill sets, a productive temperament, time, and money—a long but not exhaus-tive list of attributes and resources of a fine farmer. Your farm will need all kinds of resources at the right place and the right time, as well as an even-keeled planfulness from you, the producer. Take stock now of what *you* bring to bear on the development of your farm business, the resources available to you, and what you must learn, rent, purchase, or hire in to make a go of it.

CHAPTER REVIEW QUESTIONS

1. What relevant business experience do you bring to your farm operation?

2. Are you good with numbers?

3. What are you good at growing?

4. What sales and marketing skills do you possess?

5. What kind of leader are you? Do you have a leadership track record?

6. How will you balance perfectionism with pragmatism?

7. Do you have both time and money to get this business rolling? If you have only one of these precious resources, how will you use one to compensate for the lack of the other?

CHAPTER THREE

HARVESTING SUNLIGHT

Ellen

Now that we have asked the deeply personal questions "Why be a farmer?" and "Do you have what it takes?" and you have begun the process of honest self-appraisal, let's go back and start at the very beginning, taking nothing for granted. In other words, let's talk biology, where all the magic happens. The question now is: "What exactly is farming?" Here are some basic biological concepts that you need to consider as you start your farm—natural laws that are fundamental to the business called agriculture. These are the concepts that continue to inspire us even after decades of daily work.

THE GIFTS THAT KEEP GIVING: SUN, SOIL, AND WATER

We must never forget that the original source of energy in all of farming—and for all Earthly life—is the sun. Whether it's the daylight that your crops turn into sugar via photosynthesis or the release of energy from the breaking of chemical bonds via natural decomposition processes in the soil, the sun is the beginning of all cycles. All combined, this exquisite blend of chemistry, physics, and biology remains a modern-day magic show; even our finest scientists remain befuddled as to how this process originated. After all, no laboratory has successfully re-created life, despite infinite combinations of amino acids, electricity, and grant money. It's no exaggeration to say that the creative power of sunshine is as close as we'll ever come to pure alchemy in our lifetime.

Our job as farmers is to always maximize what's free, and first and foremost that means harvesting sunlight in the form of plants. When it comes to farming, plants—vegetables, flowers, fruit trees and bushes, nut trees, grains, pasture, and herbs—are king. Due to the fact that animals cannot digest sunlight directly, they are secondary producers who must live off the aforementioned "magic" of photosynthesis. It's worth noting that plants (along with a few microbes) are the only life forms able to live directly off the sun's energy. Only they can take this power and convert it into sugar, the building block of calories that all humans and animals need for survival. No matter what your final crop may be or become—from strawberries to cheese to tractor-trailer loads of beer—growing plants, and growing them well, is mandatory.

This means that it behooves you to have a robust understanding of how plants work and what they need to thrive. Here's an equation to study and understand:

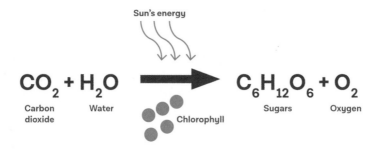

As you read the equation from left to right, carbon dioxide (CO2) and water (H2O), when exposed to sunlight within the leaf biomolecule called chlorophyll, go through a chemical reaction. The resulting products are sugars and oxygen.

Lucky for us, the supply side of this equation is all freely available: Carbon dioxide is ever-present in the air around us, and water is obtained by the plant via its roots. The plant then makes sugar, which it uses to power its metabolic processes. Even better, it gives off oxygen, crucial for all of us aerobic critters to breathe. At the most basic level, plants need lots of leaf surface to receive sunlight, as well as an abundant root structure to keep finding water and nutrients. The farmer's job is to help create the conditions that will make our plants as robust as possible, giving them every opportunity to thrive. No matter what type of farming you ultimately choose, growing plants will be your primary responsibility, and you really have to love it.

In pursuing the goal of harvesting sunlight, our secondary job as farmers is to take care of the natural resources upon which our businesses depend. Beyond sunshine, let's move on to what also nourishes our plants: the soil.

Soil provides the context in which our plants can grow, combining physical, biological, and chemical conditions in a whole underground universe. On the physical plane, soil is the medium in which plant roots take hold, which allows them to remain upright. The physical components of soil (sand, silt, clay, organic matter) comprise the necessary structure for plants to take root. This is also where the complex (and poorly understood) chemical–biological dance between plant roots and microbes takes place.

Soil particles have surface area with an electrical charge, where plant nutrients are held. These particles get together to form aggregates that act as the house for soil microbes to live in. All of these participants—the soil, the soil microbes, and the plant roots—coexist, enabling the plant to find the water and nutrients it needs to grow. Sometimes, as is the case with legumes (plants in the pea and bean family Fabaceae), the plants and microbes collaborate to sequester nitrogen from the air, placing it into the ground where any plant can use it. Nitrogen is freely available from the air. Suffice it to say that, in order for us to harness the power of the sun in the form of crops, sustainable farmers maximize what's free in order to help take care of the soil.

Okay, so now we have sunlight and soil. That's an excellent start, but we need one more key ingredient to get our plants to grow: water. Obviously, water is necessary for all life forms. Soil organisms, as well as our plants, need water to survive. But why, exactly? Plants need a steady supply of water to help move the nutrients from the soil up through the roots and into their aboveground parts. Look back at that photosynthesis equation (see page 36), and you'll find water right there at ground zero. Carbon dioxide (again, free from the air) with water (also free from the air—notice a trend?) are the ingredients with which plants make sugar. No water equals no sugar; no sugar equals no life.

Consider the water resources available to you on your farm. Mother Nature gives us water in three distinct realms: precipitation, surface water, and groundwater. What are the rainfall patterns in your area? You need to understand not only the absolute quantities but also the timing (i.e., specific months), average frequency (e.g., every day, once a month), and intensity (ranging from pounding thunderstorms to days-long drizzles). All of this weather data will impact how you farm. Surface water resources include ponds, streams, lakes, and rivers. You may be able to tap into these sources (there are rules about this in some locales), but there will be accompanying costs of pumping and piping to get the water to your fields. In most cases, groundwater is wonderfully clean and immediately usable via a well. You'll need to know how much water your well yields, and if you don't have a well, it will likely be a low five-figure investment to get one drilled and running.

Sun, soil, and water: These are nature's gifts to us. If you already have farmland, take an assessment. What do you already have to work with and how will you farm so that you enhance these resources, rather than deplete or waste them? If you don't have farmland yet, don't worry—we'll help to steer you in the right direction in chapter 5.

But farming isn't only about sun, soil, and water. Farming is a human activity—this is where *you* come into the picture. Becoming a steward of the land means learning how your biological farm works through observation, experience, and research. You need to be curious and capable. Some good habits include:

- Observe and see what is so.
- Ask "*Why* is that so?"
- Compare your knowledge and experience with that of other farmers.
- Learn and follow up through books, websites, and agricultural conferences.

Taking care of and making good use of the natural resources we are blessed with will set the parameters of our success, and observation and intellectual curiosity will go a long way toward utilizing them effectively. We will delve further into these fundamental concepts, addressing soils from the ground up (pun intended), in chapter 6.

DIVERSITY

Like all important things in life, maintaining diversity is a balancing act. For example, if you grow only one crop, you overexpose yourself to the risks of drought, floods, pestilence, and adverse market conditions. Overextend your-self by attempting to grow too many crops at once, and you risk growing none of them well. You want enough diversity, but it is possible (and more common than you might think) to diversify too much in farming. In the right measure, diversity is desirable because it equals resilience.

It's a core biological tenet that ecosystems are healthier and more sustain-able when they have a diversity of members. However, the set of conditions that favors one set of organisms may be detrimental to another set of crea-tures, and vice versa. Since conditions in nature tend to change and fluctuate over time—even over the course of twenty-four hours, and undoubtedly over the course of centuries—an ecosystem will tend to flourish if it has various diverse mem-bers that can take advantage of a fuller range of conditions. The same is true for your agri-cultural business. Philosophically, this means that the system (i.e., your farm, your life) can withstand changes in conditions without fail-ing, so long as there is diversity throughout.

For a weather-based business, this is a cru-cial concept. Without exception, your farm will undergo challenges due to natural causes. This is why experienced farmers grow some crops that prefer cooler, wetter summers, as well as varieties that enjoy hotter, drier sum-mers—you can't know which kind of summer you're going to get. In some years, late blight arrives on the wind from a Southern state, destroying my second and third plantings of tomatoes. Inexplicably, some years it doesn't. I can't know ahead of time what a given year will bring. Accordingly, I always want to grow *some* tomatoes, but not be completely reliant on tomatoes to keep my farm profit-able. So pragmatically, I want to grow some

DIVERSITY THROUGH REDUNDANCY

On Lloyd Nichols's vegetable farm in Marengo, Illinois, diversification against inclement weather is built into his annual crop plan. Because he manages four hundred acres of vegetables, sometimes the weather is sunny and warm on one end of the farm, while on the other, the edge of a thunderstorm inflicts severe damage to crops. To hedge his bets, Lloyd plants the same crops in multiple locations, creating a redundancy that greatly reduces the chances of being wiped out by a single weather event.

blight-tolerant tomato varieties, even though they are less delicious, just in case the blight does come.

Challenges also result from man-made causes: Your partner gets sick, or your worker quits. A national food-borne illness scare means that no one wants to eat fresh spinach for several months. Your favorite tillage tool breaks, and the replacement part is no longer available. Sometimes, these all hit at once. This means that you should have more than one person who can do any one job on the farm, that you'd better have more products than spinach to sell in May, and that you'll need to have more than one tillage tool on hand (or one that's easily borrowed or rented), in case yours breaks. All of these types of diversity, from purely biological to systemic—as well as resource and labor redundancy—are going to help you and your farm flex and bend, but not break, under pressure.

On the other hand, there is a pervasive notion that the perfect, diversified farm must have both plant and animal crops—that the farms we read about from the early 1900s are what we should return to. Those farms had multiple livestock species (poultry, cattle, sheep, and swine) as well as field crops (corn, wheat, oats), pasture, workhorses, and maybe even vegetable crops and a dairy. And for sure, there is a real aesthetic and emotional beauty to those systems. This, undoubtedly, is why journalists and other well-meaning farming outsiders have popularized the perceived romance of these operations. But I've frequently observed new growers attempting this level of diversity, and it rarely works out.

Consider how much we've changed as a society since the early 1900s. Those beautiful farms from long ago had multiple families living on them, with many adults contributing to the caretaking of such a diverse set of enterprises. Uncle Dave was the swine handler, Aunt Sue took care of the laying hens, Dad worked the row crops, and Cousin Sam ran the cattle and made hay.

This kind of multi-enterprise vegetable, crop, and animal farm may still be a great option now, but not without a diverse set of dedicated grown-ups to manage it. One person, or even one couple, is going to be hard-pressed to be good enough at so many different enterprises to make them economically viable. Being an expert at one craft is challenging enough; attempt to master two or three or four crafts, and the results quickly skew toward mediocrity rather than success. Therefore, *a diversity of individual skill sets must exist before a diversity of enterprises can occur.*

What I've observed of some fledgling, highly diversified farms, is that none of the enterprises are managed well enough to make them profitable—the time and talent of the individual operator is spread too thin. Unless there is a very well balanced work schedule based on lots of experience, it often happens that too many crucial tasks need to be performed at the exact same time. And, in farming, timing is everything. Additionally, tracking inputs and expenses on multiple enterprises is far more complicated than tracking those of just one, so it's seldom undertaken. As the farm begins to mature, it never becomes clear which of the many enterprises is actually making money. It's important to know which parts of your business are really paying the bills, and, in some instances, the diversity of having animals alongside vegetable, flower, or grain crops might not be a practical choice. You don't want to direct your two most limited resources, time and money, on crops that won't pay.

So what is diverse enough? Here are a few sensible ways to consider diversity for your farm business.

- Stack your seasons (e.g., spring greens, summer greens, fall greens; or spring strawberries and fall pumpkins).
- Grow more than one crop or species within an enterprise (e.g., hay and grain crops; poultry and swine; root vegetables and leafy green vegetables).
- Grow more than one variety within a species (e.g., three different kinds of cucumbers, such as picklers, Asian, and regular slicers, and then further diversify with four different varieties of slicing cucumber, such as Marketmore, Raider, Dasher, and Straight 8.
- Pursue more than one type of market channel (e.g., one farmers' market and a CSA; restaurants and an on-farm store; wholesale and retail).
- Have more than one person trained to perform each important farm job.
- Establish multiple ways to run your system (e.g., never rely exclusively on a particular tractor or specific piece of machinery for crucial daily operations).

SEASONALITY

One of the pleasures of farming is the inextricable waltz with the seasons. Of course, your geography will dictate the nature of those seasons and how different they are from each other. But no matter where you farm, there will be seasonal fluctuations in some or all of these factors: temperature, sunlight (intensity, angle, and duration), and precipitation (form, amount, intensity). For those of us in the temperate zones, frost dates are a mixed bag—sometimes welcomed, and other times cursed. Your job is to understand the seasons of your farm and find out how you might capitalize on them and mitigate their negative effects on your crops.

Between the consequences of climate change and the proliferation of products and innovative season extension techniques, most vegetable farmers now grow and sell for more weeks of the year than they did even twenty years ago. I used to only have vegetables to sell between June and October. Now, the normal selling season in my area is mid-April to Thanksgiving. It is quite common for growers here to have fresh products nine or ten months of the year, with some even attaining full year-round production. With the advent of winter farmers' markets and vibrant CSAs (community-supported agriculture), growers have the option to sell their shelf-stable wares (jams, sauces, salsas), frozen goods (meat, summer-excess berries, beans), and seasonally independent products (eggs, dairy products, bread) every week of the year.

Growing on the edges or "shoulders" of the main summer season is more work, but it can pay off. Having fresh produce at these special times usually commands a higher price. Customers are very appreciative of having more fresh food for more months of the year, and it makes your sustainably grown produce more competitive with the seasonless array of foods at the common grocery store.

However, knowing when *not* to grow crops is crucial as well. I mean this in a couple of different ways. First, I would not grow a certain crop in a certain season if I didn't believe it would grow well or, frankly, even taste good. This is the situation for growing spring broccoli in Virginia—it gets terrible flea beetles (a difficult insect to control) and matures in very warm temperatures, leading it to taste spicy and mildly unpleasant (more like a radish). So, while it *can* be grown, the quality is very often poor. The solution? Don't grow spring broccoli. It is crucial, then, to understand how different crops grow in different seasons—where despite consumer demand, it is a bad idea to try to grow them at certain times. Just because there is year-round demand for a product, such as broccoli or pastured chicken, it may not be profitable or wise to grow it every month of the year.

Second, downtime can be incredibly important for the rejuvenation of both soils and people. When I was a seasonal vegetable farmer, I relished having three months a year completely out of the field. Of course, there were still unavoidable responsibilities: paperwork and bookkeeping, ordering supplies, and meetings. But for me, December through February also had plenty of time for rest, travel, and pursuing hobbies. This time off was essential to gather up the energy to start the next season. Downtime is the energetic counterbalance to the extreme uptime of full-on spring planting season and deep-summer harvest time. Some farms, especially those with livestock, have a steadier year-round workflow. But it is still important—borderline mandatory—for these kinds of farms to operate more or less on cruise control for several weeks sprinkled throughout the year, so the farmers can take a breather.

INFRASTRUCTURE

Here's a scenario that I have encountered regularly over the years: A customer would approach me at the farmers' market and object to my prices because they grow tomatoes in their backyard, and "Tomatoes are so easy to grow. How could yours possibly be so expensive?" The first time this happened, my instinct was to justify my prices via some perfect retort, but then I realized that the reasoning behind my answer might not be obvious to the layperson. Technically, the customer was correct: Growing a tomato plant, in and of itself, is no tremendous feat. But this patron's "cheap" tomato only exists under a particular set of circumstances—the growing of a single tomato plant (or a glorified few) for his personal use. The customer has willingly invested free labor and has no other business considerations to contend with.

Namely, the home grower need not take into account the cost of infrastructure. If the price of our commercially grown tomatoes needs to cover only our labor and the seed, then indeed $4 a pound sounds outrageously high. But in order to consistently supply lots of tomatoes to lots of people, month after month, the market price must pay for the land, the trellis, the 2-hp pump in the well, and the electric bill (not only so we can operate the greenhouse, barns, and farm office but also so that we can charge our cell phones and the electric scales for market). Additional costs may include the deer fence, the gravel on the farm road that allows us to tend the patch even after a two-inch downpour, along with taxes, insurance, and the truck and fuel to get to market. The price we get for our crops has to cover every expense incurred in our operation—*every single one*—or we will not succeed. Good luck explaining that to the grumpy customer in front of you! However, for our own peace of mind, we need to understand this concept of infrastructure in our core.

Here's another common scenario: A well-meaning customer or acquaintance says, "I have some property out in the country. Maybe some nice young farmer could use it to grow [fill in the blank]." At first blush, this offer is appealingly generous. Indeed, one of the primary requirements in farming is the land itself. But there are many other investments that need to be made before a raw piece of land can be considered a farm. Each of these items falls under the heading of infrastructure and improvements:

- The physical things attached to the land: buildings, electrical service, water for irrigation and washing and processing, fencing (to keep livestock in or deer out), roadways
- Rolling stock: equipment and vehicles for production and marketing
- Processing equipment for washing, packing, and cooling your products (outdoor sinks and drain systems, freezers, walk-in coolers, and dry storage)
- Office supplies: computer, phone, internet connection, a secure (dry, rodent-proof) place to store records and permits
- Staff and customer needs: any necessary parking, housing, restrooms, handwashing area, lunch area, meeting space

Barns and Buildings

When it comes to barns and other buildings, for most producers, there is no such thing as too many roof-covered spaces on a farm. Protection from the elements is crucial for keeping produce fresh and salable, securing livestock during inclement weather, giving you and your team a place to stay dry, warm up, or get out of the beating sun, and most commonly, for keeping important things dry (fertilizer, hay, papers, seed, electronic devices, and equipment). There are many options for creating protected spaces on a farm, from simple lean-to sheds to plastic covered hoop structures, to repurposed shipping containers. All of those cost money and time to erect. At minimum, it would be swell to start with one roofed structure. Consider what structures are available to you, and determine what you might need to build or purchase.

Equipment

Beyond land and the infrastructure listed above, all agricultural enterprises need at least some equipment. Depending on the type of enterprise, this can range from a few key tools to a wide array of expensive machinery. Simply put, farming is often a materials handling game. For example, here's a short task list for moving materials, for one patch of tomatoes over the course of one season:

1. Get the tomato transplants and crew to the field for planting.
2. Get a wagonful of mulch hay to the patch.
3. Move T-posts to the rows for trellising.
4. Harvest the tomatoes and bring them back to the packing shed.
5. Load the tomatoes onto a truck to get to market.
6. After frost, retrieve the trellis stakes and put them away in the barn for the winter.

The parallels are nearly identical with animal agriculture. The farmer must move cross fencing, water troughs, feed and supplements, and ultimately get the animal to a processing facility of some kind. People learned long ago that wheels make moving materials easier. So, you will need some sets of wheels—carts, wagons, and trailers. Something has to push or pull those wheels around. The most basic power source (besides the sun!) is your body. Typically, systems become more complex as you move from people-power to horses, and then to tractors. No matter how simple or sophisticated an operation you want to run, chances are you will need some tools, equipment, and power sources beyond your own body.

Fortunately, there are many ways to acquire or gain access to perfectly good farm equipment—beyond paying a premium for new products. Here is a list of options:

Buy it outright. Try like heck to buy well-priced used equipment. It's almost never worth it to buy new—you can live without the "new-car smell." If you don't know how to shop for equipment, find a friend who does or find an independent dealer who regularly attends auctions and sales and can keep an eye out for things you need.

Borrow it. This can be hard on relationships, so tread carefully. Leave it better than you found it—oiled, greased, and clean from dirt and debris. If you break it, even the smallest piece, you must repair or replace it. Of course, trouble comes when you need it at the same time that the lender needs it, so strategize for that well in advance.

Rent it. Only some tools are available for rent, but perhaps more than you might think. Stop by different equipment dealers and see what they have. Many agricultural co-ops and NRCS (Natural Resources Conservation Service) offices also have equipment to rent. Renting is mostly appropriate for tools that you need infrequently—that is, once a season or less.

Hire a custom operator. In most agricultural areas, there are folks who can travel to your farm and perform custom work for you using their own equipment. Prime examples include combining grain, making hay, hauling livestock, some types of tilling, and seeding. This option can really save a lot of time, money, and stress. Potential issues here include timeliness (whether the operator is available when needed) and the size of the job (whether it's big enough to be worthwhile to the operator).

Outsource it. Say you need hay for your animals. You can either obtain the tractor, the haybine, the rake, and the hay baler to make it yourself (not to mention the required three consecutive days of temperate weather, or the all-important experience to learn how to produce it well, or the barn to store it all summer), or you can simply buy perfectly made hay from someone else, exactly when you need it. Let's say that you need transplants for your vegetable patch. You can either build or rent space in a greenhouse (with the aforementioned time and experience requirements), or you can contract a professional grower to custom-grow your seedlings.

Do without, by making do with what you have. Potato harvesting would certainly go better and faster with a proper root harvester, but this machine— even used—is expensive. Can you get by with using a single-shank subsoiler? Or a moldboard plow? Or a plastic mulch lifter? This option of "doing without" can be temporary or permanent, depending on how long it would take for you to pay off a piece of equipment with labor savings and possibly increased product quality. When you do the calculations, you may be surprised by the results—it may be way cheaper to get the tool when labor savings are figured in, or it may take many years to pay off.

Land is just the beginning of the infrastructure of your farm. Putting these other components in place will take careful planning, time, and possibly a fair bit of money.

LIMITING FACTORS

My time working with Gary Zimmer, a pioneering biological farmer from Wisconsin and founder of Midwestern BioAg, gave me a deep appreciation of limiting factors. This concept is a version of Liebig's Law of the Minimum, which states that growth is controlled not by the total amount of available resources, but by the scarcest resource.

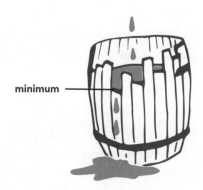

As you can see in the illustration, the barrel can only be filled as high as the lowest stave. Put into farming terms, let's postulate that the barrel represents the full picture—that is, a sustainable farm. The staves of the barrel might represent soil resource, climate, people power, equipment, financial health, markets, and cash flow. So no matter how good the weather is and how many good folks work for you, if you can't sell your products for a good price, this "lowest stave" drains your potential, and you will flounder. Or no matter how great your markets are, if you can't supply quality products in good quantities, this becomes your lowest stave, and you will suffer. Or let's say you have done everything to increase the health of your soil. It can grow crops easily and well, but you haven't invested in a good fence to protect against the deer problem in your area. The same result looms: You will struggle.

Throughout your planning, implementation, and nurturing of your farm business, keep in mind the question, "What is my limiting factor?" Which lowest stave is keeping you from attaining your goal? Also note that limiting factors often move in incremental lockstep out of necessity. For example, how can you have more yield if you don't have good soil fertility or have adequate cash flow? How can you grow retail sales without good access to customers?

It reminds us of the proverbial conundrum, "Which comes first, the chicken or the egg?" Sometimes, you must inch up your limiting factors with painstaking slowness, even though you'd rather do so quickly, installing entire barrel staves at a time. Instead of progress, it can often feel like a vicious cycle. This strong desire for rapid improvement is where new farmers often make the mistake of going into debt, trying to buy a quick solution. Don't give in to this temptation! Keep working away at inching up multiple staves at a time.

Another cautionary note: Before you get worked up about fixing a limiting factor, be sure that you know what area really needs to be worked on sooner rather than later. I've watched many a grower go down the rabbit hole of one aspect of the farm, most often soil fertility, while failing to identify the true limiting factors. They might spend hours and hours reading and researching, and, in the end, purchase a few special fertility "potions" that promise to dramatically change yield. However, they haven't done the basics, like making sure that the soil is the proper pH, or they continuously drive on wet soil, creating semi-permanent compaction, or perhaps they're trying to grow crops in the wrong hardiness zone to begin with. These farmers are focusing on icing the cake, when they don't even have the ingredients yet to bake the cake!

The challenge is to accurately assess what your true limiting factor is. That might require gaining some perspective by standing back from the problem and asking others for their take on the situation. It might even entail misidentifying your limiting factor and experiencing failure several times early on. Don't worry about these missteps; instead, consider them opportunities to learn (more on this in chapter 13). Over the course of my farming career, I've personally struggled with a dizzying number of limiting factors. They've ranged from soil fertility (hard, dead soil) to sales staff (unmotivated) and from deer fencing (ineffectual) to infrastructure (insufficient). Rest assured, all good farmers experience similar challenges.

THIS IS A BUSINESS

We've been talking about harvesting sunlight and turning it into food or fiber (along with its attendant challenges and infrastructure requirements) and ultimately selling products to a customer. If it isn't clear by now, being a farmer is being in business, a clear distinction from homesteading, hobby farming, or backyard gardening. Many folks come into farming to exercise their passions for food, nature, and meaningful work. But in order to keep having the opportunity to farm, you have to get the bills paid. Regardless of your personal motivations for wanting to farm, if you are unable to pay your creditors, your taxes, your employees, and especially yourselves, you will have your farming privileges unceremoniously revoked. Farming must be a passion. Business must be an attitude. A wise producer intentionally cultivates both.

You will be markedly more successful in farming if the idea of being a businessperson appeals to you. A businessperson has to care about quality as well as quantity, people as well as products, customer relationships as well as personal concerns. So a farmer needs to be not only a caretaker of the biological systems on the farm but also a caretaker of an entity called a business. Businesses have relationships, cycles, and needs just like living plants and animals.

When making agricultural decisions as to what, when, and even how to plant, you must simultaneously keep business principles in mind. The most basic business rule of farming is this: *Income must exceed outgo.* If costs outstrip your income, the whole thing comes to a screeching halt. While at first glance this concept might seem obvious, consider this: At the time of this writing, the average American household owes $132,000 of personal debt, $15,000 of which is in high-interest credit cards. Do you think that you'll be able to manage your personal debt on top of running a farm where bills outstrip receipts? It's a tremendous challenge, and, unfortunately, hundreds of thousands of farmers have tried and failed.

This is why "income must exceed outgo" is a rule every new farmer should strive to apply as soon as possible. Understandably, this will take time to fully manifest, as many new businesses endure up to five years of operating losses before turning a profit. Again, that is where part-time work and family loans may come into play. But setting this financial goal early on is essential for a new farmer to be able to navigate the predictable—and necessary—growing pains of gaining experience, while remaining economically solvent. It will also build a foundation of financial strength that can eventually lead to the choice of taking on sensible debt.

Let's add this important business principle to our list of foundational concepts laid out above and see how a few different scenarios might unfold.

- If your goal is to create an organic vegetable farm, but you have insufficient amounts of surface or ground water for irrigation, chances are good that at various points during any one season, your plants will languish. Yields will be compromised. Your water resource has to match your crop choice.
- Say you adopt the practices of bio-intensive farming (a methodology that uses very high plant population densities), but you don't have a rich enough soil to support such spacings. You will end up with the perfect plant population, but none of the plants will yield much, or at all. Your soil resource must match the farming method.
- If you create a business model based on being the supplier of the most unusual foods (say, for example, ginseng, shiitake mushrooms, and ginger), and all of the crops have the same heavy-labor crunch schedule, you will likely need a big crew for just a few weeks, which is difficult to pull off. Labor management and cash flow will be a struggle.
- If your business model is to provide fresh chicken when no one else does (during winter), and your winters are harsh, those chicks may not grow well, or at all. Talk about a financial loss! Sometimes, when nobody around you is growing a particular product, it's for a very good reason.
- Each of these is a situation where new farmers might easily convince themselves that they have a great innovative idea. Yet, in each case, without thinking through every detail of the plan and making sure they have all the fundamental concepts of farming covered, they will likely face poor financial results.
- Remember, you are creating a business that relies on nature, along with

your finesse and determination to harness the sun's energy. You must bring to bear the other natural resources available to you, along with an infrastructure that's uniquely suited for your enterprise. You need to incorporate diversity into your system, all the while accounting for variables of seasonality and limiting factors. And you must operate with financial discipline, treating your farm as a business on track to take in more money than it spends.

These are the hard goals that all new producers must face, and ones that will determine your farm business success and personal satisfaction.

CHAPTER REVIEW QUESTIONS

1. What water, sun, and soil resources are available at your chosen site? If you don't yet have a farm, what resources are needed for the set of enterprises you envision?

2. How will you incorporate the principle of diversity in your operation?

3. What infrastructure do you have? What infrastructure elements do you need? How would you prioritize those needs from "absolutely necessary" to "would like to have someday"?

4. What limiting factors exist in your operation? Or, if you are just getting started, what limiting factors do you anticipate?

5. What about the idea of running a business scares or excites you?

CHAPTER FOUR

YOUR $0 STUDENT LOAN EDUCATION

Ellen

How do you learn how to farm, anyway? If you think about it, that's really a crazy question. How is it that most of us don't know how to cultivate the most important resource (besides oxygen and water) for staying alive—food? Why, it's the result of progress, of course! Almost all of us have been relieved of the burden of having to grow our own sustenance. It's the blessing of modern life. Or maybe not.

Once upon a time, way back 150 years ago, we were almost all either farmers or gardeners. Most folks lived in rural areas, and the general store supplied only staple ingredients such as flour, sugar, and oil—those items that are harder to grow and process on a home scale. People in other professions, such as manufacturing, teaching, medicine, and mercantile, still grew vegetables and kept a few chickens in the backyard. Farms that grew grain, raised cattle, and produced milk also had kitchen gardens to supply vegetables and fruits for the household. As we know, those days are long gone. Gone, too, is the invisible day-to-day, week-to-week, season-to-season transfer of farming and gardening wisdom from one generation to the next.

So here we are, two or three or more generations removed from the hands-on knowledge of how to grow food. What's a wannabe farmer to do? Basically, the answer is simple: Get thee to a farm. The basics of how we learn remain intact. We learn by observing others and benefiting from their lessons, as well as by trial and error.

Student debt in the United States is at an all-time high. Even former President Obama was forty-two years old before he finished paying off his student loans. Big debts hang like albatrosses from the necks of young people. The good news for those seeking agricultural knowledge is that you can get paid while you learn, if you structure your education accordingly. You won't get rich during this learning process, but you won't walk away with any debt, either. All things considered, it's one heck of a deal.

IS THERE VALUE TO A FORMAL EDUCATION?

The short answer is yes, of course. The longer answer, however, is to weigh whether it's worth the time and expense. On the most practical level, there are very few farming programs at colleges and universities that will teach you how to farm—that is, in the sense of how to fix a broken irrigation pipe, how to adjust a vacuum seeder, or how to assist a distressed ewe in birthing mode. (Notable exceptions do exist, such as the Agroecology and Sustainable Food Systems program at the University of California, Santa Cruz, and the Sustainable Agriculture program at Central Carolina Community College.) But, there are a number of useful ideas and fundamental philosophies that you can take in from a soils, plant science, or business management class.

For me, it made sense to get my BS in agriculture, so that I would have a college degree in the hopper for whatever future endeavor might arise. But it's hard to make the case that anyone coming into agriculture post-college should head back for more expensive university coursework. If you feel the need to continue a traditional education, I would recommend finding classes at a local community college that could help round out your knowledge about biology, business, and management. There are oodles of successful farmers who have no formal agriculture coursework under their belts. Rather, upon further study, they probably either worked a number of years for other successful farmers or simply dedicated years and years of growing to eventually become good at farming.

GOING IT ALONE

There's no doubt about it, trial and error is a very solid teacher, often more valuable than a traditional university education. The problem with relying solely on the School of Hard Knocks, however, is that it can be a very long road to success, and all the while, the costs of the mistakes keep mounting up. Indeed, while some folks have figured out how to farm all on their own—with the help of some carefully studied books and instructional videos—it's not the shortest or most reliable path to success. And, as mentioned in chapter 2, you must have either lots of time or plenty of money, if not both. If you have neither the time nor the finances, you won't be able to sustain a seat-of-the-pants business plan. So, let's investigate how you can foster a zero-student-loan education by learning from those who have already found success.

ELLEN'S FARM BEGINNINGS

I got my first farm job at age sixteen. The farm was on a tiny island in the Potomac River west of Washington, DC. I rowed across the river to work each day, joining a ragtag group of hippies to hoe strawberries and pick squash. I was so young and naïve that I don't think I really learned anything about farming, per se—just about sweating profusely all day. It was, however, my first glimpse into a whole new world, nothing like the neat and tidy suburbia from which I came. This place was full of hardworking idealists, those choosing a different path from any I'd ever been exposed to at home or at school. These folks were willing to give up some creature comforts and big salaries in order to engage in tangibly meaningful work.

Thus began my journey into farming. I, a wide-eyed daughter of a federal employee and a full-time mom, a person who had never stepped foot onto a farm before and had no knowledge of any farming ancestors, turned hard left into agriculture.

While my story is uncommon, it is not unheard of. Over the last thirty years or so, most new farms are started by folks who were raised in nonfarm households. Why have farm kids stopped becoming the next generation of farmers? I suspect that it's because America's get-big-or-get-out federal agricultural policy since the 1970s has stoked intense competition to achieve economies of scale. Kids growing up on farms witnessed their parents struggling financially, physically, and emotionally to keep the family and business afloat during rough period after rough period. Those farmer parents encouraged their children to go to college and get "real" jobs elsewhere. If this was the mission, then it certainly worked. Kids left their family farms in droves.

The consequence? Not only do less than 2 percent of us actually know how farm, those who do are not always passing our knowledge down to the next generation. We are left with a fundamental disconnect: The average age of farmers in the United States is fifty-eight years old, while a high percentage of the young folks who want to farm are from the city and don't know how to do it. The brain trust is aging out, and they are not conveying their wisdom to the newcomers who are actually interested in farming. As the Apollo 13 astronaut was famously reported to have said to NASA's mission control, "Houston, we have a problem."

Lucky for us, there are a number of seasoned first-generation farmers who are committed to teaching anyone who has the gumption how to farm. These mentors, by and large, have degrees from both formal colleges and the university of life. They made all the mistakes, took all the financial risk, and have come through to the other side—to success. These are precisely the sort of folks we want to learn from. The best zero-student-loan education I ever got was from seeking out these experienced teachers and working alongside them. I kept my eyes open and my mouth closed. Then, after I had observed, pondered, and reflected on the work, I was able to ask informed questions.

I was fortunate to learn farming from a couple of seasoned growers. My first important farming teacher was Hiu Newcomb, at Potomac Vegetable Farms. She and her husband, Tony, were counterculture, idealistic first-generation farmers. I worked for her for four seasons in my teens and while in college. It was not a formal internship—I worked by the hour, and she taught me what I needed to know to do good work. Slowly, as I learned the systems and strategies, I attained higher-level production skills. As it turned out, she and her daughter Hana hired me again seven years later. This time, I was to be a salaried full-time farmer!

Over the next four years, I worked to transition the satellite farm that the Newcombs owned into a fully functioning, organic vegetable farm. Even with my several years of seasonal work behind me, this was truly the beginning of my trial-by-fire farming education. And learn I did. I had thirty acres of ground; two ancient, gasoline-powered, tricycle front-end tractors; and a pull-behind disk plow for tillage. What I didn't have was a fuel tank, fencing, irrigation, or any other equipment. It was pretty much a blank slate.

During these years, I went from conference to conference, where I'd ask farmers, agriculture professionals, and sales people for advice on transitioning those acres from conventionally grown sweet corn to organic tomatoes. I made some great progress in those first years. And then, a different nature called—my biological clock went off, and it was time for me to start a family.

Luckily for me, along came my second important farming mentor. An experienced CSA farmer named Heinz Thomet (now of Next Step Produce) just happened to be loose for a season, and was looking for a farm to land on. It made great sense to invite Heinz to farm with me, to help cover the bases while I was preoccupied with pregnancy and new motherhood. Heinz lived in a tiny apartment in my house, and we got to know each other very well. I had 24-7 access to his amazing mind—that is, when my new mommy brain had moments of peace.

In today's world of sustainable agriculture, we are blessed to have a few seasoned generational farmers, those who were born on a farm, who delight in passing on their skills and knowledge to newcomers. Heinz was one of these people. He was no back-to-the-land hippie; he was born and raised on a farm in Switzerland. He was, and is, an authentic farmer, which to me means someone who has farming ingrained in his DNA, someone who can weld, back up a wagon on the first try (pronounced "vagon" by Heinz), and whose rhythms are already synchronized to the seasons. He taught me about equipment maintenance, how to create order out of chaos, and how to quickly hook up three-point hitch equipment. There were lessons within lessons, unlocking knowledge that I had never known existed.

So in exchange for making room for another grown-up in my home, I received a one-of-a-kind education. With Heinz's guidance, we embarked on major farm improvements. We cleaned out barns and sheds packed with decades of trash and five feet of aged manure. We changed the layout of the fields to be on the contour. And after careful consideration, we purchased a few new implements that markedly changed our efficiency.

But the most educational piece was working with him to create a cohesive farm production system for the land. It was the interaction, the conversation, and the perspective that he brought to the situation that really changed the way I fundamentally understood farming. After our two years together, I could see how the equipment choices interlinked with crop timing and profitability. I had a much deeper understanding of how to break down any farm job into individual tasks, then to apply ergonomic considerations, and finally, to arrange the tasks in time and space for efficiency of movement. This education cost me nothing and gave me priceless training.

ON-FARM LEARNING MODELS

Internships and Apprenticeships

While internships and apprenticeships are, by definition, different arrangements, the terms are typically used interchangeably when applied to farm work. This is a mistake, and it's a hot-button issue for me! An apprenticeship is a formal agreement between an employer and a person who wants to both work and receive job training and education, so that they may then become a farmer. Apprenticeships typically last at least one season, sometimes two, and are paid. On the other hand, an internship is a less formal arrangement, whereby a student gains some valuable work experience. Internships are much shorter in duration, usually a few weeks or months. Interns may or may not receive wages in exchange for their work.

Given all these restrictions, why would a farm ever offer an internship or apprenticeship program? First, only individuals with a serious desire to learn farming would commit to a long and rigorous apprenticeship program. These kinds of folks can provide extremely high-quality work over time as their training begins to inform their performance. It is also a valuable avenue for grooming prospective permanent management-level employees. Most often, farms offer paid internships that include smaller educational components, with the emphasis on getting work done. For most growers who undertake either of these options, the real payback is in the satisfaction of offering such a valuable service to the next generation of growers. It's a great way to pay it forward.

Unfortunately, there are farms that incorrectly label their work positions as internships or apprenticeships. They simply teach someone the proper way to hold a shovel, how to put a seedling into a pot, or how to sell a bunch of beets at market. Showing someone just enough to actually perform a task isn't an internship; it's just being a manager of an entry-level farmworker. Both internships

KNOW YOUR RIGHTS AND RESPONSIBILITIES

The exact definition of how an intern differs from an employee is under debate in the legal system. Each state may have its own interpretation. In the meantime, these are the conditions that must be met for an intern to receive less than the minimum wage:

1. The internship is similar to training that would be given in an educational environment.
2. The internship experience is for the benefit of the intern, not the farmer.
3. The intern does not displace regular employees, but works under the close supervision of existing staff.
4. The employer that provides the training derives no immediate advantage from the activities of the intern.
5. The intern is not necessarily entitled to a job at the conclusion of the internship.

and apprenticeships need to have formal learning structures—what will be taught, in what order, using what techniques, with what kind of practical application. Often, apprenticeships provide reading lists, formal discussions, and field trips. A true apprenticeship is a golden opportunity to learn the art and craft of farming, while at the same time earning a decent wage. A good internship is a summer job that exposes a beginner to the world of agriculture.

A warning to those who are shopping for internships and apprenticeships. Watch out for farms that do not offer a formal learning component. Often, these farms just want to find people who will work for low wages in exchange for the promise of providing a so-called education. In my experience, they are generally not financially successful, and thus can't afford to pay normal, fair wages. This is not the kind of operation you want to emulate. Go to a farm that has a good track record with past interns or apprentices and takes this educational role seriously. Or, if you're not concerned about a dedicated educational element, you can always get a job as a farmworker.

Work for Wages for at Least Two Years

Farms need help, and most of them hire employees to get the work done. Get a job, and you'll get paid to learn how to farm. Same as above, do your due diligence to make sure that the farm is legitimate and well respected, and treats its workforce humanely. The best course is to meet the farmer and work a day or two before signing on for the season. Speaking to previous employees is another excellent way to make sure you know what you are getting into.

In my experience as both a worker and an employer, the second season on the same farm yields exponentially more knowledge. Having gone around the calendar of seasons once is great. But it's not until you go around the second time that many of the *whys* of what you've been taught really fall into place and start to make sense.

For example, I have always farmed using tons and tons of hay mulch for weed control. My first-year workers can't believe just how thick I want that mulch to be applied in order to accomplish the goal of totally weed-free patches. It just seems silly and wasteful to them. Oh, and it's a lot of work, too. So, they mulch light and airy. Guess what happens come July? Yup, the mulch isn't thick enough to keep sunlight from reaching the soil, and dormant weed seeds are stirred to life. Sigh. All that time and expense, and there are still weeds. We will struggle with that patch for the rest of the season.

Second-year workers, though, know that if they get the mulch thick enough, the only time we'll have to visit that field again is when it's ready to harvest. The *why* of the mulching project is now viscerally understood. This learning by repetition, and observing the results of your own work, happens best when many of the variables are stable—when you are working on the same farm, with the same soils, climate, and tools.

That's the main upside of year two on the same farm. The second benefit is that the farmer knows exactly what skills and training you've had and will trust you to handle them independently. That sets the stage for the grower to teach you the next set of skills and techniques to advance your learning. Here, your zero-student-loan education is funded by your time and your commitment to growing your skills.

Some of the most successful new farmers I know worked on other people's farms for several years before venturing out on their own. We highly recommend this path. To shorten the time frame slightly, try to work in a region that matches the climate that you'd like to finally settle in. Numerous specific farming practices and techniques, and certainly timings, can vary quite widely based on geography and weather. If you think that you'll end up back near your parents in Florida, don't learn how to farm in Duluth, Minnesota!

Beginner Farmer Training Programs

These programs are now more popular than ever. They fill a need to provide technical training to a range of folks—from total beginners to those with some growing experience. Training programs combine formal educational workshops with field trips and work assignments on local farms. Some programs are free, while others charge fees for the season. Most are part-time only, allowing the trainee to have a part-time job elsewhere. I know a number of growers with settled families and home lives who managed to learn how to farm working through this kind of program. It was the perfect fit for their schedule, because they couldn't otherwise pause their lives and move to a farm to work.

Farm Incubators

Incubators are part educational program, part business-development programs. Since the early 2000s, a plethora of farm incubators have sprung up across the country to help new farmers get some real-life, low-cost practice running their own farms, while providing them with in-house resources, all under the tutelage of a senior farmer. Incubators are typically viewed as something of a graduate program reserved for farmers who have some experience (perhaps through an internship or college courses) but want to take the next step toward independence.

Here is how a farm incubator works: The organizing entity (usually a nonprofit) secures farmland and necessary infrastructure: greenhouses, coolers, tractors, implements (machines pulled by a power source), washing stations, fencing, etc. The land is divided into plots and leased to new growers for a set fee. Use of the equipment and infrastructure are billed by usage. For example, greenhouse and cooler space are rented by the square foot, and tractors and implements are rented by the hour.

The incubator farm usually has staff that provides some training and business support. The goal is to create an affordable platform upon which a new grower can begin to build a farm business without needing to take out a loan for capital purchases. Think of it as a safe environment to test your skills and ideas. Some incubators also assist in marketing. Others even offer to pool different farmers' products together to sell under one incubator brand.

The application process is competitive and requires that the applicants have some growing experience already, as well as have a farm business plan. Incubators are a superb second stage for a new farmer to gain valuable experience and to build up a financial track record. Many growers "graduate" from the incubator to either acquire their own land or to become professional farm managers. For some growers, this is the point where they determine that farming is just not going to work for them as a career. And thank goodness that they haven't gone into tremendous debt to find that out.

OTHER OFF-FARM EDUCATIONAL AVENUES: CONFERENCES, FIELD DAYS, WORKSHOPS

I'm a conference fanatic. I love the energy, the camaraderie, and the information. As a lifelong farmer, especially an organic one, I have felt somewhat alone out in the field. Coming together with others facing similar struggles and triumphs always fills me with hope, inspiration, and technical tidbits. The sustainable agriculture community has a number of special attributes, but chief among them is the attitude of generosity—of spirit and of information. Biological or organic growers have always relied heavily on each other to learn and hone the craft of farming.

For decades, the conventional system of university-generated research, which is extended to the farming industry through the federal extension service, had very little to offer the sustainable agriculture community. For the most part, conventional researchers thought and said we were crazy. Times have indeed changed, but the best information about growing is still shared between growers. Get yourself to some conferences, sit in the front, ask the speaker out to lunch, and make connections. There's a world of tips, tricks, and basics to gather from conferencing. The costs are palatable, and many offer discounts to beginning farmers.

Some of the most important developments in my farm philosophy and practice came from listening to growers talk about their operations. For example, I heard the Nordells speak about their crop rotation system twelve years ago, and it completely changed the way that I farmed thereafter. The Nordells have been farming vegetables using only horse power for more than thirty-five years in Pennsylvania. They developed a crop-rotation strategy to combat weeds and increase soil health that they call "weed the soil, not the crop." The Martens, of Lakeview Organic Grain in upstate New York, taught me about cultivation tools and timing. From J. M. Fortier, author of *The Market Gardener*, I learned about tarping for soil preparation and weed control. Andrew Mefferd, author of *The Greenhouse and Hoophouse Grower's Handbook*, showed me how crop steering works in hoophouse production.

Smaller workshops, seminars, and field days abound. This is your chance to step foot onto someone else's farm, and see for yourself how other growers go about the business of farming. There's no telling what gems you might discover there—a better way to tie the ends of drip tape, a great knife sharpener, a new pepper variety, or even a new customer channel. Keep your eyes peeled, and take lots of photographs. Compare your crops to theirs in terms of vigor and yield. Again, these events are affordable, or even free. Pay special attention to those given by Extension or NRCS (Natural Resources Conservation Service),

TIMES HAVE CHANGED

I had the distinct honor and pleasure to become acquainted with the Virginia NRCS cropland agronomist Chris Lawrence a few years back. He and I made a few short videos about soil health from a farmer's perspective. While working on that project was amazing, what happened next forever changed my relationship with service providers. Upon learning about my crop rotation system, Chris had some great ideas about how I might achieve my soil-building goals using organic no-till techniques. We decided to conduct some trials on my farm to test the practicality of using a no-till planter to seed winter cover crops and summer green manures. On one hot summer day, there were three different government professionals—one federal, one state, one local—working with me to figure out how to use a rented seeder. Wow, I would never have believed such a collaboration was possible. Many of these folks are now dedicated to our success. This is a new age in agriculture, and now organic and biological farmers are considered valuable members of the agricultural brain trust.

as these publicly funded professionals are now keen to help our segment of agriculture. Get to know your service providers, and invite them to your farm.

DIY Farm Tours

My friend David Paulk is now in his ninth farming season. He and his wife started their farm in midlife, after Dave retired from a twenty-five-year military career. Before they took the plunge, they heavily attended conferences, took a beginner farmer training program, and visited farms all around the mid-Atlantic area. Dave didn't wait for a field day or workshop; he called people up and asked if they could visit. Through these DIY farm visits, Dave gained tremendous insight into the current state of farm technology, marketing, and labor management.

Get onto as many farms as you can and in as many ways as you can. Offer to pay or to provide some labor or service to compensate the grower for their time away from a busy day—buy a bunch of products from their store, write a great online review, pull a bunch of weeds, and be incredibly grateful. Then, like Dave, when you are really farming up a storm years later, you can provide the venue to help educate the next new farmers.

STAY-AT-HOME EDUCATION

Books are still king in my house. I have shelves of farming books, and I've even read most of them. Some treasured tomes are dog-eared—spines cracked, full of sticky notes and pencil marks. We highly recommend availing yourself of this relatively inexpensive (or even free from the library) avenue of learning. Trade journals and magazines add to the vast body of information in print. I even subscribe to *American Vegetable Grower*, which keeps me apprised of happenings on the national landscape, and in conventional Big Ag—it's smart to know what the guys at Dow, Dole, and Driscoll's are doing, too. We need to pay attention to the big picture of food and agriculture in our society, keeping ourselves educated about all aspects of production.

With YouTube's endless offerings of video instruction, it's easy to lose entire evenings to the promises of a computer screen. But with a little digging, some valuable information is easily available for free. Social media sometimes provides a useful forum for picking up good production information, as well. A number of very good growers are sharing great photos with informative comments full of technical details. Again, just watch out for getting lost in the ether. Farmers need a solid eight hours of restorative sleep.

Another way to learn is by hiring a professional to perform a task and then watching them work. Yes, you're paying them to perform a service, but the educational component is free. Almost all of my mechanical knowledge came from appreciatively observing a hired mechanic attack and solve my problems. I asked questions and stored those answers away for the next time I found myself in a bind. The same holds true for plumbers, electricians, carpenters, welders, or masons—if there's a job to be hired out, chances are 100 percent that you can glean valuable knowledge simply by observing and asking some questions.

Better yet, when you cultivate and maintain good relationships with these hired masters, they can end up coaching you over the phone. I'll never forget how each time I called my contact at the local dealership about a tractor that wouldn't start, he'd ask, "You got fire in the wire?" He was asking if the battery was grounded, and whether there was enough spark to engage the starter solenoid. From him, I learned how to troubleshoot my way through a series of diagnostics, eventually landing at the reason why my old International Harvester wasn't turning over. By the end of the call, my tractor would be back up and running. This is a valuable result for the price of a free phone call.

The road to mastery comes one hour, one mistake, one lesson learned at a time. It's as if the Ten-Thousand-Hour Rule were written specifically for farming—repetition begets prowess, muscle memory, and stamina. There are no shortcuts. But unlike our traditional education system, where students rack up huge financial debts on their way to getting their degrees, farmers have the opportunity to get paid to learn both the art and science of farming. And once you've gained years of valuable experience on someone else's dime, you can continue your education through reading, watching, and listening to growers, researchers, and wise teachers. Thus, the road to farming know-how is potentially free; it just requires time, intentionality, and persistent attention.

CHAPTER REVIEW QUESTIONS

1. What avenues of education will you take to learn the art and craft of farming?

2. How much time do you have available to pursue your farm education?

3. Do you think a farm incubator program might be a good fit for your farming education?

CHAPTER FIVE

TO BUY OR NOT TO BUY— GETTING YOUR HANDS ON LAND

Forrest

"**H**ow can a new farmer afford to buy land these days?"

This is, without a doubt, the most common question I receive when I speak in public about farming. Whenever it's asked, heads across the audience swivel, and ears perk up. It's a riddle that everyone wants to solve. After all, what's a farmer without land? It's like a fisherman without an ocean, or a pilot without a sky. A farmer, by definition, requires a farm.

Yet, over the years, I've gradually come to realize that this is the wrong question entirely—and not because the answer itself isn't important, even foundational. The issue of buying and affording farmland is, arguably, one of the greatest challenges of our time, and certainly of our collective future. And it's precisely due to the gravitas of the challenge that I always give the same response: "They can't. In fact, I've rarely met a full-time farmer who could afford his land, even if he already owns his farm."

Of course, this isn't the answer most people are expecting or can immediately wrap their minds around. I can empathize; I was once in the same boat. But it's an honest response. Over my career, I've witnessed scores of farmers forced to abandon their dreams for the simple fact that they didn't fully grasp the ramifications of land ownership until it was too late. Because of this, knowing how to ask the right questions about getting your hands on land could be the first and greatest challenge for a new farmer. Your long-term success might actually depend on it.

Let me illustrate this conundrum with a true story. Many years ago, I was proudly telling my old college roommate, Chris, how, after decades of financial difficulty, I had finally turned my family's farm around and made a profit. All of the bills were paid, there was no outstanding debt, and I had even managed to squirrel away a modest salary for myself. Chris, who had recently graduated from one of the most prestigious MBA programs in the country, listened intently as I spoke. When I finished my humble brag, he asked me a straightforward question: "How do you compensate for the cost of the land?"

This was an easy one, or so I thought at the time. "I inherited the property," I said matter-of-factly. "So I don't have to account for it."

He shook his head, disagreeing. "You *own* the land. But every day that you hold onto it without getting a return, you're *paying* for it. How do you pass this cost on to the customer?"

I assumed he must have misunderstood me—or perhaps he was just confused about how this whole farming thing worked. My friend was a lifelong city boy, so I decided that I should give him the benefit of the doubt. "You're not hearing me," I tried again. "The land is already paid for. There's no mortgage and no debt. So there's no cost."

Chris now regarded me as though I was a bit batty. "Forrest," he said, "think about what you're saying from a business perspective. How much would your farm be worth today, from a real estate standpoint?"

I considered this and replied with a fair-market valuation.

He nodded his assent. "So you'd agree that if you suddenly sold the farm tomorrow, this would be a reasonable sales price?"

"Sure," I replied. "But I don't ever plan to sell."

"Stay in the hypothetical for a second. For the sake of argument, what if you woke up tomorrow with no farm but instead had all of that money. What would you do with it?"

I replied without hesitation. "I'd invest it, of course."

"So you wouldn't let that cash just sit around. You'd want it to generate an annual percentage return, right?"

"I'd be crazy not to," I said. "Annual inflation alone would cause it to depreciate."

"True. So, just to be perfectly clear, you'd expect a return on your investment and not just long-term capital appreciation—but also some extra interest, dividends, or rent, as the case may be, along the way."

"Yes."

"Sounds very reasonable," he agreed. "But now you realize the problem, right? You've been doing the exact *opposite* when it comes to your farm."

I still couldn't make the connection. "No," I said, remaining adamant. "I'm debt-free, and I'm charging enough to make a profit. . . ."

Chris studied me, his expression quizzical. "You're making a profit on your *operating* costs," he insisted, "not on your land investment."

"But I didn't *pay* for the land. . . ."

"But don't you see, you *are* paying for it. Every day that you don't receive a return on the land," he explained, "is a day you could be doing something else with that money. Instead, you've convinced yourself this is a 'free' asset, and you're subsidizing it with your time—or your philanthropy, or whatever reason you might choose—for a zero percent return."

I must have still appeared confused, because he decided to try a different tack.

"Think of it this way," he suggested. "Say you had to start over tomorrow from scratch. Could you do it?"

I was forced to shake my head. "No. There's no way I could afford the land."

"Exactly. But why?"

I shrugged, at a loss. "Because I don't have enough money."

"Yes. And the reason you don't have enough money is because you haven't factored full replacement costs into your prices—in other words, you haven't included a return on the investment in land." He paused. "Forrest, think like an entrepreneur for a second. Would anyone invest hundreds of thousands of dollars—even millions—if all it did was generate a profit above operating expenses? Of course not. Like anyone else, they'd expect a reasonable return for their original investment and have a way to pass this cost along to the customer."

I made a final, feeble attempt to counter his logic. "But . . . I could access that money any time I needed, right? I'd only have to . . . to . . ."

"Sell the farm?"

And just like that, he had me in checkmate. "Yeah."

My friend nodded. "That's the only way I can see your making a true return—by selling the land. But if you do, you go out of business." His eyes were solemn. "Forrest, that's *not* a sustainable business model."

I'll be the first to admit that I'm pretty thickheaded. But when this logic finally sank in, it was suddenly all so clear, so obvious. In fact, I had already seen this puzzle play out dozens of times before among local farmers who had been forced to sell their land due to revenues that couldn't cover the full extent of their bills. The simple fact was that these landowners had never factored in a full price of their cost of production—and this had to include a return on their land investment (or replacement cost), even when they owned it free and clear.

Would a factory not account for the cost of its property, a dentist the expense of her office building, or a baker the price of his rent? Of course not. Economics 101 tells us that we must pass these costs along to the customer. No one invests money without expectation of a reasonable return. But in reality, this is precisely what I—as well as a huge percentage of other American farmers—was doing every single day.

As a result, farmers like me were growing food at artificially cheap prices and subsidizing the bottom line with our own goodwill. But when times got tough, we also bore the full economic brunt. And because the commodity system (the vast market where I started out selling my crops and where the overwhelming majority of farmers still do) provides essentially fixed prices, even if producers *want* to charge more, it's incredibly difficult to do so. Hence, for most producers, selling the land often becomes the only viable option for recouping the true value of their farms.

This realization was both shocking and heartbreaking, and I went round and round with it in my mind for many years. But after that revelatory conversation with my friend, I vowed that I'd do everything I could to stop this from happening, not only for my farm, but for others, as well. The book you're now holding is part of that promise.

THE CONSEQUENCES OF CHEAP FOOD

So why should you, the new farmer, care about any of this? Because, either directly or indirectly, you'll be affected by this bizarre accounting practice. Ironically, when landowning farmers don't fully charge for their cost of production, it creates a sprint to a food-pricing bottom, where only large-scale producers are able to achieve the most stable financial results. This is because the commodity system rewards maximum efficiency—that is, how much a farmer can squeeze out of his land and equipment. And the larger the farm, the greater the economies of scale. Maximum production is also a determining factor in how many dollars in government subsidies (payments in the form of agricultural price supports) a farmer can receive, helping bolster his bottom line. All combined, it's a self-perpetuating mechanism of artificially cheap food, driving the cost of production far below where it should be. (For more about the commodity system, see chapter 7.)

Hence, a domino effect occurs. If your next-door neighbor doesn't value the replacement cost of his land, then you can't either. Or if your neighbor buys newer, more efficient equipment or purchases the latest variety of high-production genetically modified seeds, then you're suddenly at a competitive disadvantage. It's a race to the bottom, but straight up a mountain of expenses. Like many aspects of contemporary farming, it's a paradox that makes little sense whatsoever.

It gets worse. Over the decades, acclimated to these artificially low prices, consumers have been trained to think that food really should be this cheap. Consequently, it's now nearly impossible for farmers to charge fair, full prices for their products, even if they want to. When food prices are raised even a little, it will often make headlines on the national evening news.

If you're looking for evidence, it's all around us. Have you ever noticed that we've had the exact same 99-cent hamburgers, the same 99-cent chicken burritos, the same 99-cent french fries coast to coast for close to three decades? As of this writing, I can get a pizza delivered to my house for $5.99, the same price as back in 1995. That's insane.

On top of all this, throw in the variable of increasing farmland prices (often because of new home construction and commercial development, where land becomes valued for lawns, not agriculture), and the facts become starkly black-and-white. All combined, these factors make it nearly impossible for a new farmer to afford to buy land—at least on any sort of large scale.

But don't worry, and don't be discouraged. We have a whole basketful of farmland options for you in the second half of this chapter. Remember, this book is called *Start Your Farm*, not *Forget Your Farm*! Before moving forward, however, it's worth devoting another page or two to understanding how and why this financial fiasco happened to begin with. After all, those who don't study history are destined to repeat it.

THE FASCINATING HISTORY OF FREE FARMLAND

So why in the world wouldn't a farmer account for the investment price of his property? The answer is actually quite simple. It's primarily due to millions upon millions of acres of land giveaways, farms that were handed out scot-free for hundreds of years. That's right, *free land*—much of which has subsequently been passed along with minimal cost from generation to generation to this day.

The history of North American land giveaways started as early as the sixteenth century, when enormous parcels were distributed for the use of farming. Later, after the Revolutionary War, many soldiers were given acreage in lieu of paychecks. Around the same time, Spain was giving away land in Florida to anyone who would farm it, and Mexico was doing the same in California. When they became states, this land was grandfathered in. And in the largest land giveaway in American history, the Homestead Act of 1862, 270 million acres were offered free to 1.6 million farmers, a practice that continued until 1976 on the continental United States (and until 1986 in Alaska).

Over the years, the allocation of free acreage increased from 160 to 320 to as much as 640 acres. Who here could use 640 free acres of land to start their farm?

It's widely accepted that there are 900 million farmable acres in the United States. This means that with the Homestead Act alone, nearly a third of America's farms were handed out at essentially no land-investment cost to the producer (some of this was open-range and mountain land, which was quickly divided into pastures with the invention of barbed wire in the 1870s).

In the short term, this helped feed a hungry, growing country. But it also created huge numbers of farmers and their progeny who, through inheritance, never had to account for the fair-market investment value of their properties.

Thus, generation after generation—and to this day, continuing right before our eyes—America slowly established a system that doesn't fairly value the true cost of food production.

Weird? Absolutely. Accurate? You'd better believe it.

AFFORDABLE FARMLAND IS A TWENTIETH-CENTURY CONCEPT

What about those millions of other farmers who *have* bought land, as practically everyone who owns a farm but didn't inherit it has had to do? After all, plenty of farmers in the past made a tidy living off their acreage. Documents show that my own ancestors, who purchased the farm in 1816 with five thousand dollars, enjoyed many excellent years far into the twentieth century, generating enough revenue to exceed their operating costs and return on land investment combined. As recently as the 1950s, my grandfather bought additional farmland and turned a profit. Why, then, is it so hard to repeat this accomplishment today?

There's a clear explanation here, too. Prior to 1970, there was actually a long period of time when farmland was priced at a rate where crop prices exceeded the cost of production—even with a return on land investment properly factored in. Classically, this was what we now recognize as the traditional family farm, diversified operations that had multiple generations working on them. For the better part of a century, millions of producers made an actual living buying a farm and working the land.

Things really flew off the rails, however, around 1975, and these two charts from the USDA show why. Rising land prices, combined with declining commodity prices (note how corn, soybeans, and wheat steadily retreated in a nearly inverse direction relative to increasing land prices), dictated that farmers suddenly either had to produce more, become more efficient, somehow absorb and endure the lower prices, or, as previously noted, sell their land and cash out.

It's easy to see that when commodity prices fall, there's hardly any middle ground. During this period, hundreds of thousands of small and midsize traditional producers simply tried to ride it out, hoping to stay afloat through a period of ever-falling agriculture prices. Eventually, however, the accounting error of never pricing for the investment cost of land finally caught up with

US Farm Real Estate Values, 1950-2011*

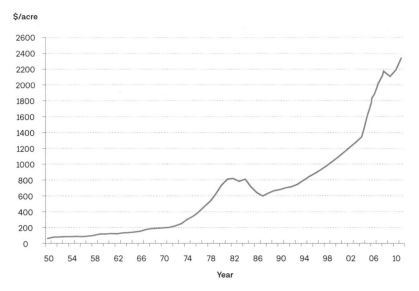

$/acre

*Includes all land, buildings, and dwellings on farms.
Sources: USDA, NASS

Inflation-adjusted corn, wheat, and soybean prices, 1912-2014

Index, 1940 = 100

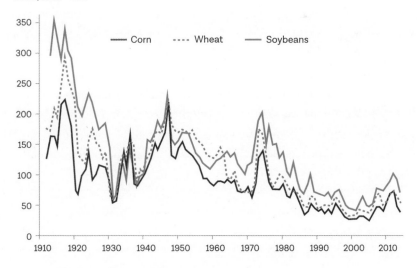

Sources: USDA, Economic Research Service calculations using data from USDA, National Agricultural Statistics Service and US Department of Labor, Bureau of Labor Statistics.

them, and farmers couldn't make up the shortfall simply by working harder or producing more. Those who had loans and mortgages were especially devastated. This was called the 1980s Farm Crisis, when hundreds of thousands of family farms went bankrupt.

So who makes a living today in agriculture? As you might expect, large-scale producers—those with many thousands of acres of crops, many units of animals, or some other form of high-volume production—make up the lion's share of this demographic. These farms are also the ones that receive the most federal subsidies, with 20 billion dollars of annual payments. According to the USDA, out of a total of 2.2 million American farms, 840,000 of them depend on subsidies to help survive from one year to the next. This translates to nearly 40 percent of our farms, a staggering number! For good or ill, this is what happens when farmers commit to selling into a commodity-based market, one that doesn't compensate for full production costs. It's worth repeating that, even if farmers *wanted* to change, the way this system is now so entrenched, it would be enormously difficult to do so.

Yet, there's another kind of producer making a living, as well. This is a different type of farmer entirely, one who insists on accounting for *all* investment costs, and has figured out a way to pass along an authentic price for the cost of production. These producers seek out their own markets. They grow food on a scale that suits their acreage, climate, geography, and personal abilities, regardless of what politicians or policy-makers suggest they should. They connect directly with their communities, forging relationships that buoy them through economic turmoil.

These principles form the core of sustainable agriculture. This is what we believe in. This is what we want to share with you.

This brief history of farming economics should completely change our perspective on that original question. So, instead of "How can a new farmer afford to buy land these days?" it would be far more constructive to ask:

- "How can a new farmer gain access to land at a minimal cost?"
- "How can a new farmer charge appropriate prices, ones that reflect the cost of land investment?"
- "How can a new farmer do both simultaneously?"

Ah, now here are some questions that we can work with! Let's dive right in.

AMERICAN AGRICULTURE IN THE EARLY TWENTY-FIRST CENTURY

No one—and that means *no one*—gets anywhere without lots of help from others. Farmers are no different. Traditionally, older farmers tutored the young, as their elders did before them. Wisdom was shared, methodology instilled. These folks knew everyone in their communities—the guy who could weld the manure spreader back together, the crew who pruned the apples every other year, and the old lady who helped pluck turkeys each autumn. Having elders on the farm provided the emotional support of knowing that someone had been there before and had made it through the rough times. One doesn't need to be sentimental to understand the deep pragmatism inherent in this multi-generational, help-based system.

Naturally, times change. As we've noted earlier, fewer people than ever are choosing a career in farming, and the average age of the American farmer is fifty-eight years old. By definition, this means that our pool of agricultural wisdom is shrinking in real time. But, just as important for our purposes as new farmers, there is currently a massive shift (more than 10 percent of farms nationwide) in land ownership underway, as older producers retire from farming. If history is any measure, many of these farmers are likely to sell some of the land and use this capital for their retirement.

To put this in perspective, just between 2017 and 2020, approximately 93 million acres will change hands. If you're trying to imagine what 93 million acres looks like, consider California. As in, nearly the entire state. So who will take over the management of these farms? A recent survey by the USDA (see the chart on page 76) sheds light on this important opportunity. If we study the smaller circle, a cursory takeaway might be that only 23 percent of this 93 million acres (by the way, "only 23 percent" translates to 21 million acres, roughly the size of Indiana) will be sold outside of the family. But there's far more to glean here. In many cases, farmers will only sell a portion of their properties for retirement. What about all the remaining land, the acreage that's going into trusts or wills, that fabled no-cost farmland being passed along to a yet another generation, but one that has demonstrated little interest for farming?

Land in farms expected to transfer in 2015-2019

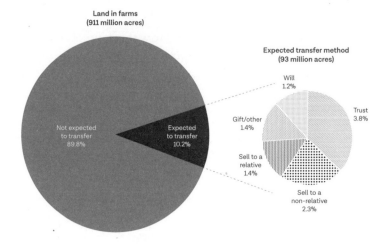

Land in farms (911 million acres)

Not expected to transfer 89.8%

Expected to transfer 10.2%

Expected transfer method (93 million acres)

Will 1.2%

Trust 3.8%

Gift/other 1.4%

Sell to a relative 1.4%

Sell to a non-relative 2.3%

Note: Data exclude Alaska and Hawaii. Percentages appearing in the smaller pie chart do not sum to 10.2 percent due to rounding. Source: USDA, Economic Research Service and National Agricultural Statistics Service, 2014 Tenure, Ownership, and Transition of Agricultural Land (TOTAL) survey.

The agricultural landscape is changing, literally and figuratively. Below are strategies for using these shifts to everyone's advantage.

ACCESS, NOT ACQUISITION

The key to success in the twenty-first century, we strongly believe, is access to land, not necessarily acquisition. We're using the word *access* as an intentionally broad term here. For our purposes, it means to have some level of production and management control on a specific piece of land without necessarily having ownership. This is a vital distinction, because if properly structured and executed, having *access* to land can often circumvent the multiple potential pitfalls of *acquiring* land—which include the upfront investment, debt, fixed location, capitalization of infrastructure, and generating that ever-difficult annual return on the purchase price of the land itself.

In the long run, what truly matters most is building your experience and learning how to grow food profitably. Yes, you'll need access to land to accomplish this. But if you stop to reassess what's truly important to farming success, then in this scenario, actual land ownership could come way down the line—if ever. Talk about having your priorities in order!

This is why our biggest recommendation for a new farmer is to focus on some form of access to land, at least at first. Better yet, if you can combine this with an opportunity to grow your experience, working with a mentor who knows the land, the seasons, and the markets, then you've dramatically increased the chances of eventual success. Again, as discussed in chapter 4, gaining experience—and by this we mean hands-on, pragmatic, repetitive, enterprise-specific experience—is absolutely vital for success. In the long run, this will be every bit as important as finding land and putting down permanent roots.

To help you identify the best options for your circumstances, we've grouped our strategies into three categories: accessing land, with mentorship; accessing land, solo; and finally, creative ways of buying land.

Accessing Land, with Mentorship

Mentoring with Farmers Who Own Land
There is a new and steadily growing wave of established producers willing to collaborate with younger farmers. Hundreds of successful examples already exist, and the arrangements take practically all shapes. This surge has largely been facilitated through intermediaries—government agencies, not-for-profits, and volunteers—who recognize the critical cultural need to make this connection. With unprecedented numbers of older farmers looking to retire, the timing here couldn't be more important.

Services include websites to connect interested parties (see Resources and Recommended Reading on page 240), regional meet-ups, and conferences. National Young Farmers Coalition has done a huge amount of work on these fronts, and your local Extension service can probably help with connections here, too. In short, no one's saying that you have to drive up some stranger's lane and ask to work with them. An incredible amount of effort has already been expended to identify and link interested parties, with an ever-growing number of successful precedents.

An arrangement can manifest itself in many ways. Depending on one's skill level, a new farmer may be hired on for a year or two as a paid employee, to learn alongside the experienced producer. Then, after trust has been established (an all-important variable), an agreement toward renting, leasing, or even some variety of rent-to-own arrangement can be forged. Or perhaps the new farmer arrives with a preestablished skill set and gradually takes over an existing operation, trading work for equity. Most commonly, the older farmer creates a mutually agreeable rental contract, and the rest of the details are specifically negotiated.

The opportunity here is to leverage the expertise of a farmer who already knows the soil, the seasons, the surrounding community, and the markets. None of these advantages should be underestimated. Furthermore, having the working support of a mentor or partner—especially one who has already shouldered the burden of owning property—allows a new farmer to fully focus on growing their production skills, and perhaps even their brand. This sets the stage for the future.

Long-Term Leases with a Like-Minded Landowner

This is one of our favorite models, one that has also experienced much success nationwide. Perhaps not by accident, it has been so successful because it actually requires the producer to take rent into consideration of their prices!

Here, the likelihood of working with a nonfarming landowner is typically greater than working with an established farmer, but the mission remains the same: finding a landowner who wants to provide land access to a young farmer, without the new farmer incurring the expense of purchasing the property. And remember that—right now—a record amount of "free" farmland is passing on to new, younger landowners, most of whom have no interest in becoming farmers themselves. This could translate into an unprecedented opportunity for collaboration.

In a long-term lease arrangement, the new farmer is guaranteed a certain duration of time—typically five to ten years, but sometimes even as long as fifty years—to build his enterprise. Unlike linking directly with a mentoring farmer, however, long-term leases are typically more hands-off. The landowner is essentially a like-minded ally, one who facilitates a favorable lease structure. In this way, the relationship is often more of a financial mentorship than an agricultural one.

Depending on the experience (and personality) of the new producer, this can be the best of both worlds: land expensed at a pre-negotiated, predictable rate, but the freedom to build an independent enterprise. Better yet, with the landowner essentially subsidizing the "free" investment cost of the property, the lease is typically much lower than a mortgage payment would be.

Often the land has been in some sort of past agricultural production (especially in cases of inheritance), so barns, equipment, and other forms of infrastructure might already exist. Alternatively, the land might be completely open and devoid of improvements. Hence, special considerations might need to be spelled out before both parties are comfortable and before any new construction or significant alterations to the land can begin.

Again, there is a large number of successful precedents for this model, so with a little research, it's easy to find and adopt a system that addresses housing, insurance, and taxation concerns. If you can think up a potential problem here, we guarantee that some lawyer has already thought of it in advance. And best of all, if things aren't working out, the lease can always be broken—a business agreement that's much easier to terminate, emotionally, than if you're working shoulder-to-shoulder with a farmer each day.

Accessing Land, Solo

Renting or Leasing Compared to Buying
On average, cropland rents for less than $100 per acre, annually, and pasture for even less than that. That's an incredible discount relative to a purchase price of $3,000 per acre (the national average as of the time of this writing, according to the USDA), especially considering that the purchase price soars the closer the location is to cities. Even when the cost of infrastructure and other improvements are factored in, the value of renting or leasing, compared to buying, can be very compelling.

Renting or Leasing Without a Farm Mentor or Like-Minded Landowner
Of course, nothing's stopping you from accessing land without a mentor, a landowner ally, or a community safety net—in other words, securing a conventional rental or lease agreement. In fact, there are certain instances when

flying solo might even be preferable to working with someone. Emotions run high in farming, and expectations from a landlord or mentoring farmer can be fickle. We've seen our share of partnerships fizzle out over disagreements that might have seemed petty to outsiders but were enough to sabotage otherwise perfectly good working relationships.

That said, it is sometimes harder to find a normal type of farm lease or rental arrangement than one of the less conventional options of working with a mentor. Due to long-established relationships, conventional lease and rental agreements can be tied up for decades—especially to those previously mentioned large-scale farmers who are seeking to maintain their enormous economies of scale. Though many acreages might already be locked into long-term contracts, not all of them will be. The more you ask and put the word out, the greater your chances of finding land that isn't already under contract. Place an online ad, network on social media, and read the classifieds in your local paper (remember, many older farmers—the ones who own most of the land—still read the paper). Keep turning over stones until you discover the right solution for you.

Being creative and looking beyond the mainstream options is critical. One especially inspirational model is D-Town Farm, on the edge of Detroit. Instead of seeking land in a rural area, community leaders approached the city about a lease on underutilized parkland. The result? They were leased a seven-acre parcel of land for ten years, at the rate of one dollar per year. Now *that* is a way to access land at an affordable price! A little farther south, in Cuyahoga Valley National Park, Ohio, in an effort to preserve historic farmland while incentivizing new farmers, the community offers up to sixty-year leases—in essence, access to land for life. The opportunities are out there, and more are emerging all the time. Keep an open mind, stay flexible, and be persistent.

Free Land

Free land still exists. Unlikely as it seems, indeed, land giveaways still occur today . . . if you know where to find them. Certain areas of the country, in an attempt to stimulate their local economies and boost populations, have offered up free plots of land. This is especially true in sections of the Midwest (most abundantly in Kansas, Nebraska, Minnesota, and Iowa), where the loss of manufacturing jobs has left homes and land either abandoned or extremely affordable. Towns have assumed the deeds to these properties, and, to encourage growth, they now offer a pathway to free ownership.

The land might be free, but lunch never is. Applicants must meet certain requirements, including the commitment to remain on the land for a certain number of years. Also, keep in mind that you'll probably be facing a complete tabula rasa—a blank slate with no infrastructure, perhaps no access to equipment, and land that is probably more suited for a house foundation than for growing crops. For a budding farmer, however, this could be a perfect fit—and again, the price tag is certainly compelling.

No Land

The best way to access affordable land? Don't have any land at all!

No, I haven't been drinking, nor am I off my rocker. But why own, rent, or lease land if you don't have to?

This was the thinking of the Texas Honeybee Guild, which has built a successful agribusiness around winged, airborne harvesters in downtown Dallas. Leveraging the goodwill of city businesses and residents, Brandon and Susan Pollard place scores of hives across the city, producing honey not for the cost of a mortgage, but for their time.

Another solid option is to follow Ellen's path. When, at age twenty-eight, she couldn't figure out how to access land herself, she went back to the good folks at Potomac Vegetable Farms. Soon after, they not only hired her to manage one of their farms, they also built her a house, and incentivized her with a potential track to ownership. She worked her way into ownership by year ten, and remained on for another fifteen years after that. When she was ready to retire, she was financially remunerated for her share, and the land was returned to the farm as though nothing had happened.

As we've mentioned, there are plenty of farmers today reaching retirement age with no children in the wings who want to take over the operation. Ellen's advice? Go get yourself adopted!

Buying Land

All of this being said, there's nothing inherently wrong with buying farmland—as explained earlier, it's just incredibly difficult to do so and make a true profit. Obviously, people still purchase farms all the time. There are many reasons new farmers convince themselves that they need to own land. They want to ensure full control over production and methodology without someone looking over their shoulder; they want to know that if they improve the land, that someone won't swoop in after all their hard work and render their efforts null; they want to prove to themselves that they can stake a truly independent claim, and make a bona fide go of it. These are all very understandable and logical motivations.

Yet, for the most part, we couldn't disagree more—at least at the start. So if you're a new farmer who doesn't already own land, then first strive to identify one of the free-land or land-access opportunities listed above. But if you're already a landowner looking to become a farmer, then the biggest challenge will be achieving a sustainable return on your investment. On top of that, whether you already own land or not, you must *pay yourself for your time*, as well as budget for a return on your operating expenses (lots more on this in chapter 9).

That's some checklist! But it's not impossible. As with many things on a sustainable farm, it's often not the system that's at fault, but more likely, it's the *execution*.

If you decide that you simply have to buy your own land, then you must strategize a way to pay as little as possible for good acreage. This is a twenty-first-century paradigm shift. Recall that thousands upon thousands of other farmers acquired their properties through land grants or inheritances. This puts you at a competitive disadvantage from the get-go. And if you borrow money, you'll be saddled with paying debt on top of all that.

With all of this firmly in mind, here's a list of farmland acquisition strategies that you might not have considered—ones that could help level the playing field in an otherwise very difficult game.

Special Loans

If you're still insistent on buying land by going into debt—and to be crystal clear, for the new farmer, debt-financed land purchasing is our least-recommended option—then you're going to need a special type of loan specifically structured for agriculture. Do yourself a favor and avoid conventional banks, for two important reasons. First, even if a commercial lender is willing to take you on, most banks are geared toward large-scale, commodity-oriented farms—operations with predictable annual income. This would typically be large-scale farms with guaranteed subsidy payments or market floors. Second, specific federal lending programs already exist for new and underserved (women, veterans, minorities) farmers, as well as private lenders specializing in agricultural loans. Always take every advantage you can get.

Federal loans, often supplied through the Farm Service Agency (FSA), offer a wide spectrum of financing options: farm operating loans, micro loans, youth loans, minority and women farmer and rancher loans, and beginning farmer and rancher loans. Since its inception in 1930, the FSA has loaned more than 60 billion dollars to more than 3.7 million farmers.

Additionally, there are emerging private-sector companies that specifically focus on agricultural loans for new and established farmers. An early adopter in this area is AgAmerica, which offers long-term, fixed-rate financing for all sizes and types of farms, including land intended for forestry, agritourism, and even solar and wind farms. (See "Farming Websites and Podcasts" on page 240.)

Goodwill Funders

Going a step further, there is a groundswell of corporate "socially responsible investing," which uses outside capital to fund sustainable and organic farming initiatives.

Leading the way is a group called Iroquois Valley Farms, which raises private capital to purchase farmland, then gives farmers the opportunity to lease this land with a track toward ownership, or directly finances a mortgage. Philosophically, the land is treated as "stock" and owners become part of a comprehensive membership. In a similar way, Dirt Capital Partners facilitates both farmland transitions and long-term leases with existing farms, working with new farmers toward a pathway to ownership.

Crowdsourcing is another way to secure financing, via the goodwill of the community. Since 2010, I've observed dozens of new farmers successfully launch Kickstarter and similar campaigns to crowdsource capital from their communities. This method can help finance everything from equipment and infrastructure to the purchase of farmland. If pitched correctly (for example, offering your community members an incentive to participate, such as a discounted five-year price on fresh produce), your campaign will not only build goodwill but also simultaneously invest your customers in the success of your business. Win-win.

Conservation Easement Strategies

For larger acreages, conservation easements are an emerging opportunity. A conservation easement occurs when a landowner relinquishes the right to build additional houses on the property, trading this for a greatly reduced tax assessment—and often a one-time cash payment from either the government or an outside philanthropic party. Philosophically, most of these landowners want to see the property remain in open space in perpetuity (contracts are often for one hundred years), thus guaranteeing large blocks of acreage, regardless of encroaching development.

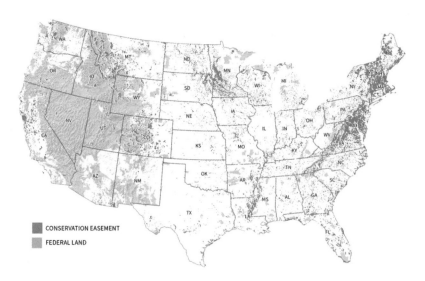

From the National Conservation Easement Database, used with permission from Bill Lane Center for the American West, Stanford University.

This has been a rapidly growing phenomenon, and we believe expanding supply will eventually create a price advantage for farmers. Due to the inability of a future owner to build houses on the property—including subsequent generations who might inherit the land—options beyond agricultural use are constrained. Thus, as more and more of these properties come on the market, it's possible that the values will begin to drop, relative to the surrounding real estate market. There's an irony here, because neighboring real estate typically increases in value due to the proximity of the property in easement; it seems that everyone wants to live *next* to a farm, just not *on* one! This is a situation worth monitoring, especially near suburban areas with their built-in access to a local customer base.

Another clever way to utilize this system is to carefully structure a deal where land is first purchased, then quickly flipped into a paid easement. As mentioned above, both government agencies and not-for-profit groups some-times offer a significant sum to offset the expense of surrendering developmen-tal rights, and these cash payments can frequently represent 30 percent of the assessed value of the land. This is a very nice discount off the price tag, if the farmer is patient and economically resilient enough to wait out what can be a lengthy process (frequently a year). These policies can vary widely from state to state, so due diligence is paramount.

Tax Liens

Finally, one of our most creative ideas might not only score you free acreage but also actually ends up *paying* you. Sound too good to be true? Believe it or not, state tax liens do precisely that: provide a guaranteed, high return on your investment, with the potential of ultimately acquiring land at no cost.

A tax lien occurs when—after being repeatedly notified—an owner fails to pay taxes on his or her property. At a certain point (deadlines to pay vary by state), this tax bill is auctioned off to a bidder (let's say it's you), who pays the delinquent sum. The deed to the property is then held by the government until the original owner pays back the taxes. However, the owner must now pay the tax bill *plus* a percentage rate of return (to you, the bidder), a number that escalates the longer the process continues—commonly, twelve to eighteen months. If, after this period passes, the original property owner still doesn't pay, then the property ultimately transfers to you as compensation. A pretty good reward for paying someone's taxes!

So why does this all happen? Each party has its motivations. The state sim-ply wants steady tax revenue and isn't in the business of managing personal property; hence, the auction. As for the owner, taxes can go unpaid for hundreds of legitimate reasons; thus, the long window of opportunity and repeated public notices to rectify the problem. For the bidder, it's an opportunity of a guaranteed return on investment (repayment of the tax, plus interest) or potentially owning a property for the price of picking up the tax bill.

Suffice it to say, this could translate into a phenomenal opportunity to acquire farmland at next-to-nothing prices, while at minimum, guaranteeing a financial return. But again, due diligence is required. One must thoroughly investigate a property—in person, before placing a bid—to make sure that a potential asset isn't actually going to be a liability. For example, it's a bad idea to farm on top of a former nuclear waste dump! Like conservation easements, tax lien policies can vary widely, and they only exist in certain states, but the list covers some of the finest agricultural land in the country.

There are so many legitimate, affordable paths to land access, it almost dispels the assumed need for land acquisition. Just remember, the land will always be out there, somewhere. Be patient, and leverage your money wisely. We can't overemphasize the need to keep financial ambitions in check, while properly growing your experience, your brand, and your passion. The onus, as usual, is squarely on you, but the solutions are out there.

CHAPTER REVIEW QUESTIONS

1. Is the thought of not owning land from the get-go a deal breaker for you? Why?

2. Do you know someone who owns more than ten acres, whether the land is farmed or not? How does he or she manage this resource? Where does the money come from to keep this acreage the way it is?

3. Imagine your dream farm's size in acreage, as well its ideal location. Using these parameters, research farm prices in your area, and determine how many fit into your category. Next, apply the principles from this chapter and see how they might enhance land affordability.

CHAPTER SIX

EXPLORING THE WONDERFUL WORLD OF SOIL

Ellen

Have you already secured access to land for farming? Now you either have to match the land to a suitable use, or you have to find land that matches the type of farm business you want to manage.

All land is not created equal in the eyes of a farmer. Climate, soil type, soil depth, and topography all directly determine what kinds of farming can happen successfully. Farms with lots of hills, or with grades over 5 to 10 percent, will not be suitable for regular cultivation. Acreage like this needs to stay in permanent cover: forested or converted to hayfields or pasture for animal agriculture. Steeply sloped land will not grow tomatoes and lettuce on a commercially farmable scale, nor will land with soil that is only three inches deep on top of rock, no matter how flat it is.

I've seen folks attempt to grow vegetables on land accessed in a variety of ways, only to find that the soil was simply not suitable for cultivation; the topography was too hilly, or the fields eroded down to nothing. This is an impossible setup, and sometimes a literal uphill battle.

You just can't get this part wrong. You must match your kind of farming to what the land has to offer, or you must find other land. An examination of what soil actually is and how it functions will help you understand why.

SOIL COMPOSITION: WHAT IS THE SOIL ACTUALLY MADE OF?
As discussed in chapter 3, agriculture is the human management of three basic

resources—soil, water, and sun—to grow food, flowers, and fiber. Land provides both the geographical setting for farming as well as the soil base from which plants grow. One might observe that humans have spent much more time and energy trying to fathom the heavens above than the ground we stand on. Many books have been written about soils, and we strongly recommend reading one or more of those listed in Resources and Recommended Reading on page 240. But for now, let's take a quick tour of the top few inches of soil that comprise our farmland.

Remember the three-legged stool metaphor from chapter 1 (page 15), where sustainability can be understood in terms of environment, energy, and economics? We can retool that metaphor to help us grapple with soil's complicated nature. In this instance, imagine physics, chemistry, and biology making up the three legs. Of course, the soil is a dynamic, complex system in which these three aspects interplay and continually affect each other, but considering them separately can help you grasp the concepts.

The pie chart below illustrates the physical content of any given volume of ideal soil—a teaspoon, a shovelful, or a dump truck load.

So half the soil is matter that you can feel with your fingers—mineral particles and organic matter. The other half is air and water, which is sometimes referred to as porosity. Take a moment to ponder these facts alone. When you dig into the earth, it isn't obvious that half of what is in the shovel is air and water, because they are thoroughly dispersed at a microscopic scale. Often, the only air we can "see" is an earthworm channel. While invisible, air molecules throughout are what keep crop roots and soil microbes happily oxygenated.

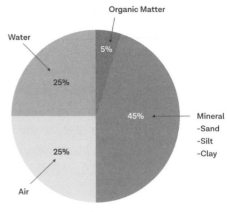

The mineral particle component of soil comes from degraded rock. Rocks break down through the physical pummeling of wind and rain, rivers endlessly flowing across them, and eons of freezing and thawing. They also degrade via chemical reactions resulting from biology living on or near them, as in the case of lichen—or more observably, a chicken eating granite grit and passing it through its gizzard. The rock pieces keep weathering, becoming smaller and smaller, eventually becoming the sand, silt, and clay that make up soil. The illustration below shows the relative sizes of these soil particles.

The exact ratio of these tiny, weathered fragments will determine the physical nature of any given soil. Soil is classified by this ratio of sand, silt, and clay, which yields descriptive terminology like "sandy clay loam" or "silt loam."

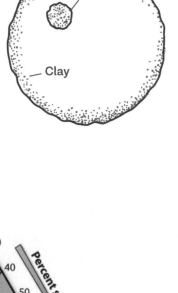

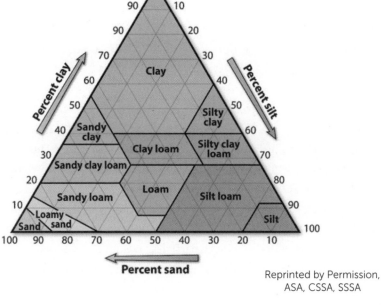

Reprinted by Permission, ASA, CSSA, SSSA

You know from any trip to the ocean what sand is—angular and plenty big enough to see each grain. To a farmer, this large particle size has its pluses and minuses. Sandy ground drains quickly, making it dry enough to drive on much more quickly than clay soil. This quality makes keeping up with soil-disturbing jobs such as tilling and weeding much easier. This same characteristic, however, makes it hard to keep plants well watered. And sand can't hold any nutrients—it's essentially little glass beads. The overwhelming majority of sand is composed of quartz (silicon dioxide), which is nonreactive and inert. So this means that farmers with sandy soils need to fertilize and irrigate more often than farmers on heavier ground—soils with more clay in them.

At the tiny end of the soil-particle spectrum sits the microscopically small speck of clay. Billions of aluminosilicate particles all stacked on top of one another create so-called heavy ground, soils that drain much more slowly than their sandy counterparts. But, due to their particular structure and chemistry, clay particles can hold onto plant nutrients *and* water. So if you farm clayey soils well, your crops will have better access to water and nutrients. The most important thing to remember about clay-based soils is that you have to stay off them when they are wet. Remember, bricks are made from clay that's had all the air squeezed out of it. Thus, one of the absolute never-nevers of farming: Don't work heavy, wet ground! Turning your soil into brick means that your growing days are over.

The third soil particle, happily in the middle of the size spectrum, is silt. Just the word itself is pleasant, silky, and delicious sounding. Indeed, it's a sweet spot for farming. Any soil type including the word "silt" is welcome news to a grower's ears. Just like Goldilocks said, this soil type is "just right." Silty soils drain well, while also holding onto enough water and nutrients for your plants. They also "work up" or are easily tilled, allowing you to use less horsepower or muscle to accomplish the work of farming.

No matter where you choose to farm, every soil on Earth is made up of some combination of these three particles. The land you farm on will forever have its own fundamental physical profile, regardless of how you might later enhance its nutrient levels and fertility. The only way to change it significantly is to truck in thousands upon thousands of pounds of sand, silt, or clay, which is cost prohibitive for anything larger than a backyard garden.

The final "stuff" of soil is called organic matter. It's that last tiny pie slice, but an incredibly important one. Soil organic matter (SOM) is sometimes called soil carbon, and carbon is the chemical building block of life. Carbon is the main avenue by which chemical energy is transported in soils.

Think of SOM like this: It is made up of the living (microbes, roots, worms), the newly dead (freshly fallen leaves, recently deceased microbes), and the very dead (rotten leaves, dead roots, manure, and dead animals—including microscopic insects and invertebrates—that have more fully decomposed). SOM makes the whole system work, bringing together mineral particles with microbial life to create the amazing, life-giving world of soil.

Soil organic matter, then, is both a physical substance (the biological life itself) *and* an important chemical force in soils—it is what ties the three legs of the stool together. The more you have of it, the better your plants will grow and the easier your work as a farmer will be. As the soil life increases, so does the soil's ability to:

- drain well *and* hold onto water
- hold nutrients
- hold air for plants and microbes to breathe (plants breathe through their roots)
- aggregate nicely, creating greater tilth (a desirable soft, crumbly quality)
- remain in place (not erode)

Did you know that organic matter is the reason why the term *organic* got attached to natural farming practices in the first place? Talk about honoring our roots! No two ways about it: Organic matter holds a key to your success, helping to unlock the potential of your soil.

Now that you have a basic understanding of what soil is physically made of, let's take a tour of the chemical realm. This will help you to better understand soil fertility, fertilizers, and how to sustainably build and balance the two.

SOIL CHEMISTRY

Cation Exchange Capacity

The best measure of a soil's ability to hold nutrients is the cation exchange capacity (CEC). A CEC determination is part of any basic report from a soil-testing lab. In the soil, many plant nutrients take the form of cations—positively charged ions (e.g., calcium, magnesium, and potassium). Silt, clay, and humus particles, conversely, are mostly negatively charged, which attract and hold those useful cations. Anions—negatively charged ions—are either not held at all (as in the case of some forms of nitrogen, boron, and sulfur) or held through complex chemical mechanisms (as in the case of phosphorus). These anions, which are quite mobile, are nutrients that should be added on an annual basis.

Simply put, CEC measures how many cations a soil can hold. Soil scientists call the place where nutrients are held and traded the exchange complex. Imagine the soil's CEC as the size of the fuel tank in your car. If the tank holds three gallons, you'll need to stop at the gas station frequently; if it holds twenty gallons, you can go a lot farther. Similarly, if a soil's CEC is three, in order for the crop plant to have continual access to nutrients, you will have to either add amendments frequently or use slow-release fertilizers. The CEC range of most agricultural land in the United States is between three and forty.

Knowing the CEC of your soil is vital to determining how and when to add amendments. Generally speaking, the soil's CEC is primarily determined by its sand, silt, and clay content, and as already mentioned, in most agricultural settings, you can't really change that—it's prohibitively expensive to add balancing tons of sand, silt, or clay. For example, to increase the sand content of a soil by 10 percent, you would need to add two hundred thousand pounds, or one hundred tons, of sand per acre, at a cost of about $2,500, not including spreading. In a small, confined space, like a hoop house or a permanent raised bed, you might be able to afford adding enough material to change the physical makeup of the soil. But out in the field, you have to banish the dream of significantly raising the soil's CEC—it's just not likely to move much in your lifetime. Instead, you have to learn how to live with and manage the soil you're on.

What size is *your* fuel tank?

Folks in the Midwest enjoy gorgeous prairie soils with CECs in the teens and twenties. Metaphorically speaking, these soils have large fuel tanks, meaning that they can take fertility additions in larger doses and less frequently. Nutrients that are applied before seeds are planted will still be there months later as the crops reach maturity.

Growers on shallower, poorer soils, however, have CECs below ten. Truly sandy soils have CECs of three to five. Those fuel tanks are relatively small. This means that applying large amounts of quickly available fertilizer (i.e., most traditional chemical fertilizers) in advance of planting is wasteful—the soil simply does not have the chemical capacity to hang onto all those nutrients. They will move down through the soil, and out of the plants' reach.

Instead, you either have to use small doses of highly available fertilizers multiple times throughout the season, or use fertilizers that will slowly and gradually release their nutrients, such as compost, rock phosphate, and other naturally occurring mined materials. At the same time, you need to improve the organic matter component of these soils. The more SOM you build, the more plant food the soil can hold. Remember, too, that decaying organic matter will release nutrients into the soil, so that's a dual return for your investment—increasing nutrient capacity *and* content!

Soil pH

Soil pH is defined as the power of hydrogen and is the negative logarithm of the concentration of hydrogen in the soil. Now, there's a mouthful. Basically, it's the measure of how many of the cation exchange sites (on the CEC) are filled with hydrogen. The more hydrogen, the more acidic the soil. The important thing to know about the logarithm part of the pH scale is that the difference between a pH of 6.0 and 7.0 is not one point, it is ten points—the 6.0 soil is ten times more acidic than the 7.0 soil.

On the exchange complex, hydrogen does not feed your crops—they get adequate hydrogen from the air and water. Instead, those hydrogen atoms in the soil are occupying spots where you'd rather have nutrients such as calcium or magnesium, so the *effective* CEC is reduced.

Soil pH affects the exchangeability of all plant nutrients. An essentially neutral pH of between 6.8 and 7.2 (true neutral is 7.0) is the sweet spot, where

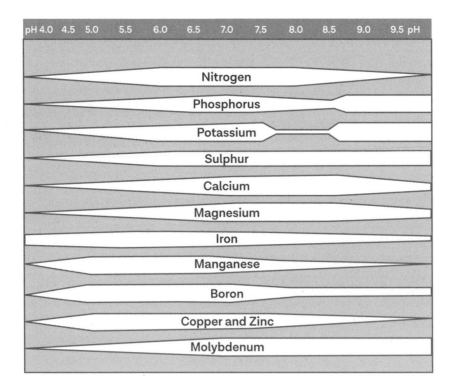

the most nutrients are the most available. When the soil is very acidic (below 6.5) or very alkaline (above 7.5), many plant nutrients become "tied up" in the soil—still present but chemically unavailable to plant roots. Conversely, in pHs outside of the sweet-spot range, certain metals are overly available and can be toxic to plants and animals alike. For all of these reasons, you must balance the soil to a neutral pH (except for the few crops that thrive in acidic soils, such as blueberries and potatoes).

The application of lime (calcium carbonate, or $CaCO3$) is the remedy for acidic soils—those below the neutral pH of 7.0. Lime is simply limestone or chalk that's been pulverized to the point of a fine, spreadable powder so it can perform its neutralizing magic.

Chemically, here's how liming actually works: In an acidic soil, the carbonate ($CO3$) separates from the calcium, reacts with the hydrogen, and forms water ($H2O$) and carbon dioxide ($CO2$). The calcium ion trades places with the hydrogen and gets onto the exchange complex. It's the carbonate that's really making things happen by removing excess hydrogen. If we use dolomitic lime ($CaCO3$ and $MgCO3$), then both calcium and magnesium will take those open spaces on the complex.

Figure 1. How lime neutralizes acidic soil

Acidic soil + Lime = Neutral Clay + Water / Carbon Dioxide / Aluminum Oxide

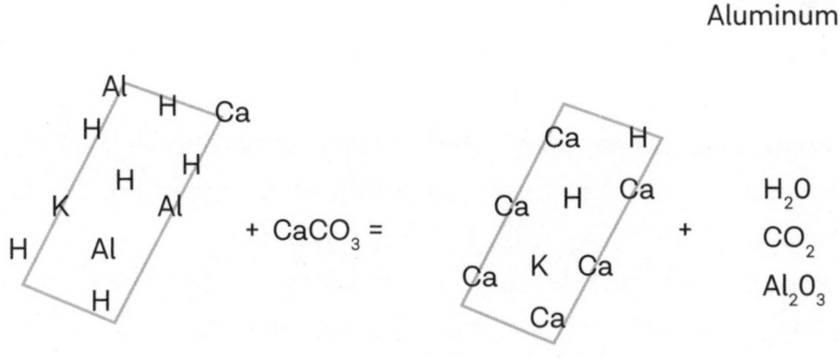

Lime is typically applied in many hundreds of pounds, up to two tons per acre. Your local county extension agents can lend a hand when calculating lime rates. Extension agents are public servants whose job it is to "extend" the

research and information generated at our state land grant agricultural universities to farmers and producers.

Conversely, the remedy for alkaline (or basic) soils is elemental sulfur. Soil bacteria will convert it to sulfuric acid over time, which will acidify the soil. Never use more than five hundred pounds per acre of elemental sulfur per year.

It's important to get soil pH right. And, it's not without cost. As you consider different sites for your farm, include the cost of adjusting the pH to your budget.

Nutrients and Fertilizers

Left on its own, the earth seems to grow plants just fine. That's comforting, but what may not be immediately obvious is that, in an undisturbed natural system, there is a closed loop of nutrients. Left alone, all of the biological products of plants and animals more or less stay inside the ecosystem. This is how our soils were built to begin with—eons upon eons of returning nutrients back to where they came from.

Then, along came modern agriculture, and the whole cycle got whacky. Today, we grow food (plants and animals) and export all of that abundance away from the farm. The problem is, unlike nature's relatively closed-loop ecosystem, we almost never bring those nutrients back in the form of waste products. The two products that result from feeding people are human waste and human bodies. Neither of these ever make their way back to the farm that grew the food. Human effluent gets flushed "away" into waterways or captured, treated, and applied to range, forest, or, sometimes, hay land. This broken cycle means farms become net exporters of nutrients. This is a fundamental conundrum for society, to be remedied another day. In the meantime, knowing that nutrients are flowing out of your farm is essential to learning how to manage the nutrients available for crop production.

Commercial growers can't just sit back and let nature take its course completely—recall, by definition, that isn't farming. You don't just hope that any old plants will grow—you plant carefully selected, specialized varieties of species, ones that your customers or your livestock want to eat. In many instances, these species would not naturally occur in your chosen ecosystem. The fundamental point is that you need to make sure that your crops, as finicky or exotic as they might be, get what they need to succeed. An understanding of fertilizer is vital.

Is Fertilizer Bad?

Confusion and misinformation abound regarding the relative healthfulness or harmfulness of almost all aspects of agriculture. On social media and in the press, topics of agricultural debate range from GMOs (genetically modified organisms) and noenicotinoids (neurotoxic insecticides) to hydroponics and animal welfare. Often lumped into the hopper of what is wrong with conventional Big Ag is the idea of fertilizer of any kind. And that is to our collective detriment.

Now that you understand the issue of nutrient flow, you can see that you must balance your soil by importing some nutrients. That is what fertilizer is—nutrients imported to support plant production. So, no, fertilizer is not necessarily bad. The determi-

WHAT ABOUT OVERCONCENTRATIONS OF NUTRIENTS?

While underconcentrations of nutrients often occur on farms, overconcentrations of nutrients occur in the conventional livestock system in the United States, with its own problematic results. Before agriculture became highly mechanized and specialized, most farms grew some animals for their own use, as well as for sale. Now, most chickens, pigs, and beef are grown in concentrated animal feeding operations (CAFOs). These systems import vast quantities of feed (i.e., nutrients). The end result is tremendous quantities of manure, concentrated in very small areas. The soils of certain rural sections of the country, especially near processing facilities (like for chickens near the Chesapeake Bay watershed or hogs in Iowa), have become oversaturated with animal manures, resulting in pollution into streams and rivers, and aerobic dead zones (where algae blooms block sunlight) in bays.

nation of whether a fertilizer is good or bad depends on what it is exactly, how it is used, and in what amount. These are details that most casual consumers don't care to ponder, and so the public dialogue gets stuck there. There is no doubt that fertilizers are to blame for polluting our waterways, but that doesn't mean that all fertilizers are bad. Good old natural manure can be a huge pollution source—in the form of urea—just as surely as chemical nitrogen can.

What matters is which products are used, and the timing and amount of their application.

The federally defined word *organic* serves to label what is allowed or not allowed in organic food production. The USDA's National Organic Program (NOP) determines which products and substances are allowed. The base rule is that fertilizers must be natural and not synthetic . . . except there are a few synthetics allowed. See how deep we can get into the weeds on this? If you want to pursue organic certification, you may want to begin your education on the NOP website.

For those who don't choose organic certification, and still want to be the best biological stewards of the land, the question becomes one not purely about synthetic versus natural but also about the effect of any fertilizer on soil biology. This is an even more difficult pursuit because there is no official list of which substances are good or bad. Thus, each of us must find our own way down this path. I chose to follow the wisdom of soils expert Gary Zimmer of Midwestern BioAg. From his perspective, some chemical fertilizers that would be prohibited for federally approved organic production are actually just fine from a soil biology angle and could be very helpful soil additions (ammonium sulfate or monoammonium phosphate, for example).

There are certainly some very common chemical synthetic fertilizers that are harsh on soil biology, such as anhydrous ammonia and triple superphosphate. As sustainable or biological farmers, we would surely avoid these. But in large doses, almost any fertilizer (including manure) could be toxic to soil life or the environment. So you take on an important responsibility as you become purchasers and managers of fertilizer.

Ellen's Journey to Soil Fertility

Here's a true confession about my relationship with fertilizer. I spent the early part of my farming career believing that all the conversations about fertilizer among my colleagues was just a lot of pretending to understand what was really going on in the field. All those calculations about how much of any one nutrient was being removed by a crop—and thus what needed to be restored with fertilizer—just seemed like superfluous math to me. And as an avid consumer of organic and natural foods, I was averse to any products called "fertilizer" at all.

I chose instead to put all my effort into creating a biologically active soil, making and using excellent compost, religiously growing cover crops over the winter, and establishing soil-building crops between vegetable crops. I figured that was plenty good enough. I firmly believed that if I created a healthy soil and worked on building soil organic matter, the plants would take care of themselves. And that strategy worked really well for many years.

Then, I observed that our increase in yields hit a wall. It finally became clear to me that all that soil biology I was fostering could not create potassium, phosphorus, magnesium, calcium, or trace minerals out of thin air. Sure, biological activity wears down the rock in the soil, but very, very slowly. It struck me that if there were nutrients missing from my soil and from my compost, then they were never going to be magically manifested by soil biology. (Except in the case of the nitrogen-fixing soil bacteria associated with the plant family Fabaceae, or legumes, where symbiotic bacteria latch themselves to the plant's roots, and literally take nitrogen from the air and fix it into a plant-available form.) I now saw that if there was not enough phosphorus in my soil, just growing more and more plants was not going to generate new phosphorus. So, my new goal became abundance, not strict self-sufficiency. My farm had been working well, but I wanted it to be even better, even more productive. I had to import specific nutrients in the form of fertilizer.

Soil Correction Versus Crop Fertilizer

I received soils training from Gary, who introduced me to the idea that soil correction and crop fertilizer were two distinctly different projects, each requiring evaluating and addressing. (If you want to dig deeper into these topics, I highly recommend his books.) The core concept is that soil correction is about making deep, long-lasting changes to your soil.

The main concerns are almost always pH (see page 94) and phosphorus levels. Phosphorus is the second most needed plant nutrient, in terms of actual pounds—second only to nitrogen, and just above potassium. Phosphorus is crucial for plant cell growth and energy transfers. Phosphorus is included in the discussion of soil correction because, as a mineral, it is fairly immobile in the soil—meaning, once there is plenty in the topsoil, it will not disappear quickly, unless it leaves via erosion. Deficiencies in phosphorus need to be addressed through the use of mined rock phosphate (these rocks are naturally high in phosphorus), compost, or manure. These amendments break down slowly, so the benefits are reaped over many years.

Once these soil corrections are made, you need to assess your annual crop fertilizer program. Crop fertilizer addresses the immediate needs of the current crop. It is not about attempting to fix deep deficiencies, but rather about making sure that the current plants can find what they need when they need it. The ingredients of crop fertilizers need to range from immediate release to slow release, with the goal of total availability over the course of the growing season (very slow release materials are too slow for this purpose).

Of course, soil correction, crop fertilizer planning, and their subsequent execution must be based on what the soil actually contains. This means taking soil samples and sending them to a lab. I recommend testing every year, until the soil no longer needs major correction and has "evened out." At that point, testing every three years will suffice. Choose a soils lab and stick with it. Different labs may use different testing methods, so it's preferable to keep working with the same company over time to track your soil's progress. Many state agricultural universities offer soil-testing services. In my experience, while more expensive, private labs tend to provide more services, as well as better presentation of the results.

Now how to interpret the results? Some labs will indicate a desired range of nutrient content (in pounds per acre or parts per million), or they will indicate a level (high, medium, low), so you can see where nutrient levels are presently versus where they should be for optimum plant growth. Sometimes, the lab will make recommendations of what amendments should be added.

Unfortunately, most commercial labs only recommend traditional chemical fertilizers. Most of the time, that's not going to work for biological farmers. It's best to find a biological soils consultant or educate yourself further to figure out what fertilizer amendments you need, and in what amounts. Every situation will be completely unique, even farms that are directly adjacent to one another, so a customized treatment plan is really the best option. Take this as an invitation to learn more about soils from books, conferences, and webinars. It's actually fascinating stuff!

WHAT ARE CHEMICAL FERTILIZERS?

Chemical (aka synthetic) fertilizers are products that are refined from natural materials or completely manufactured. A lot can be said against using synthetic or chemical fertilizers. Simply put, some require tremendous energy to manufacture, and result in products that are often unfriendly to soil microbes. Beyond that, most chemical fertilizers are immediately released upon spreading, resulting in massive losses to the larger environment—the soils, plants, and microbes just can't handle that level of input all at once. Where do those nutrients go? They leach into the groundwater or run off via erosion into gulfs, rivers, and bays, where they wreak havoc on those aquatic environments. Thus, it is our contention that most chemical fertilizers have no permanent place on a biological or sustainable farm.

Nutrients and Their Functions

When you look at a bag of fertilizer, you'll often see three big numbers like this: 10-10-10. This indicates the percentage of the big three nutrients: nitrogen, phosphorus, and potassium (N-P-K), contained by volume. So a fifty-pound bag contains five pounds of each, and the rest is made up of carriers, which can be positive (if they are sulfate), or negative (if they are chloride). Most organic fertilizers also contain a large amount of carbon, which is also good.

You can probably guess that there is a lot more to growing healthy plants than addressing only three nutrients! While N-P-K are important nutrients, they are certainly not all your plants will need. The soil testing lab I use reports back on phosphorus, potassium, magnesium, calcium, sodium, sulfur, and the trace nutrients iron, manganese, boron, zinc, and copper. This is a much more complete list. Beyond these eleven minerals, there are still more that we don't regularly test—nutrients that are needed in very small quantities. Below is a basic guide to plant nutrition, so you can appreciate the roles that minerals play in plant growth and development.

One of the hallmarks of organic fertilizers is that they come from natural materials, as opposed to synthetically derived processes. Lucky for us, these natural materials are *not* pure. This means that we get all kinds of little gifts of

NUTRIENT	FUNCTION
Nitrogen (N)	Protein production for growth
Phosphorus (P)	Cell division, energy transfer processes
Potassium (K)	Control of stomata, enzymatic reactions, movement of water, sugar, proteins throughout
Calcium (Ca)	Cell wall structure
Magnesium (Mg)	Part of chlorophyll structure
Sulfure (S)	Amino acid synthesis
Boron (B)	Cell wall formation, sugar metabolism, and transport
Copper (Cu)	Metabolism of nitrogen and sugars
Iron (Fe)	Chlorophyll synthesis
Manganese (Mn)	Aides in photosynthesis
Molybdenum (Mo)	Enzymes formation for nitrogen use, and nitrogen fixation by legumes

various trace minerals, along with the big ones that we're spending the money to attain. The so-called impurities of unrefined rock phosphate or K-Mag (potassium magnesium sulfate)—or manure, for that matter—are those lucky gifts. In addition, we recommend using products that are sourced from the ocean, which contain various super trace elements that are not found easily elsewhere. This is the role of fish, seaweed, and seawater products—to bring just a hint of ocean magic to our farms.

Nutrient Balancing

While there is still some debate about the merits of balancing soil nutrients, you can use the following broad ranges as targets when addressing fertility. The idea is that, beyond the literal parts-per-million content of each nutrient, there is a desirable ratio of nutrients. Any good soil test will list these as "base saturation percentages." A good productive soil should have base saturations around these percentages:

70–75% calcium

12–16% magnesium

3–5% potassium

The contention of nutrient-balancing practitioners is that when these ratios are achieved, the plants will have the best opportunity to get what they need, when they need it (assuming good soil biological activity).

Putting It All Together

Simply put, test and balance the soil—don't add what there's plenty of, and do add what's missing. It's important to point out that a soil test is only a tool to understand the chemical realm of the soil; and even then, the test is only a snapshot of that particular farm field at that exact moment during the season. It's a detailed guidepost, if you will, and you should use it to make management decisions. What the soil test *won't* tell you is anything about the biological or physical condition of the ground. It can't measure biological activity, or indicate the tilth, aggregation, or compaction of soils.

We encourage you to get friendly with the chemistry of your soil. Take some soil samples. Get someone to help you interpret them. Enact a plan to do major soil corrections (adjusting pH and phosphorus levels), if necessary. Come up with an annual crop fertilizer mix that is biologically friendly, works well with the CEC of the soil, and fits your budget. Use it, and watch how well the crops grow and yield.

SOIL BIOLOGY

We've saved the best part for last: how the critters in the soil play their parts. This is the historic strong suit of the organic or biological grower—appreciating the crucial role that soil microbes perform in growing great crops.

Thanks to the work of pioneering soil microbiologists, we now have just the faintest idea of who these soil critters are and how they function. While there are new species of soil bacteria discovered almost daily, you need don't need to know the specific names of any of these players.

WHAT'S IN A TEASPOON OF HEALTHY SOIL?

- Bacteria—100 million to 1 billion, comprised of up to 10,000 species
- Fungi—4 miles of fungal hyphae
- Protozoa (single-celled animals)—100,000
- Microarthropods (tiny spiders and springtails)—200,000
- Nematodes (microscopic worms)—100

You must, though, appreciate that they are the conduit through which your plants are fed.

Remember, plants harness the energy of the sun to produce sugar through the process of photosynthesis. Most plants then send around half of that photosynthate (sugar) down to the root system. The roots use that energy to power metabolic functions. But, they also exude—or sweat out—a significant portion of that sugar directly into the soil. Stop and ponder that for a moment. What do you think would happen if you took half of your salary and tossed it out onto your front yard, a little bit at a time, every day? Why, you'd soon have *a lot* of friends, I'd say! Plants use that sugar to keep billions of microbes very close by, and they trade that sugar for services. Soil biology provides the plant with nutrients, water, and security services (protection from "bad" microbes), in exchange for their little piece of sunlight in the form of sugar. I imagine it as the floor of the New York Stock Exchange on a busy day—a tremendous amount of activity, with trades flying back and forth constantly.

Soil fungi are especially good at playing fetch for their plant friends. They form a fungal network that moves water and nutrients from near and far, right into the plants roots (some fungi actually live partially inside *and* outside the plant root!). At the same time, this complex matrix of fungal "string" and its sticky by-product, glomalin, helps to bind together the sand, silt, and clay particles of our soil. This is what we mean by soil aggregation—the organization of particles into discrete packages. Soil aggregation maintains pores in the soil, airways that allow oxygen and water to flow to roots and microbes. It all magically works together, so that each participant is both servant and master.

Soil biology also feeds off of decomposing organic matter. Remember that all life forms are based on carbon—after any living thing dies, the soil microbes feed on that residue, breaking the chemical bonds between those carbon chains, thus releasing energy and nutrients. This is yet another form of sunshine captured possibly years ago. If this idea seems strange to you, compare it to how we burn oil to get heat or run our engines. Coal and oil are ancient forms of dead plant matter that release energy when burned. The same thing happens with the microbial decomposition of plants and animal bodies (as the Nobel Prize-winning physicist Richard Feynman put it, fire is sunshine that's being released from chemical storage). And thus, we deepen our commitment to continually adding organic matter to our soils. In order to steadily move toward sustainability, we must keep feeding the decomposition machine that is soil biology. If you build it, they (the microbes) will come!

Our job as the stewards of this amazing system is to keep it alive, well, and balanced by:

- using natural and biologically friendly fertilizers
- *not* using toxic pesticides
- using tillage wisely and as little as possible to maintain soil aggregates and not chop up the fungal network
- adding organic matter in the form of plant residues, mulches, manures, and compost
- keeping the ground covered as often as possible with living plants or organic mulches.

There you have it, a very brief tour of the topsoil under your feet—a still mostly mysterious universe that we absolutely depend on. A healthy soil is the seamless integration of its physical, chemical, and biological components. Civilizations have fallen by ruining their soils. Farmers must heed that lesson. Our commitment to maintaining and improving the health of our soils guides us in all our farm production principles and practices. There's no way around it. The world will not get fed through hydroponics in glass houses. The living soil is what allows us to exist on this fine Earth. Our first job as farmers is caring for this most basic, precious asset.

CHAPTER REVIEW QUESTIONS

1. What is the soil actually made of?

2. What is the role of organic matter in soils?

3. How does the CEC of your soil affect your management practices?

4. Why do you need fertilizer?

5. Why do soil microbes matter to the functioning of your soil?

CHAPTER SEVEN

TO MARKET, TO MARKET!
PART ONE

Forrest

In a culture where grocery stores, fast-food restaurants, and convenience marts are ubiquitous, stocked with seemingly endless supplies of food, it's instructive to think about where all these products actually come from, how they get there, and even, perhaps, how the farmer gets paid for their work.

Think about an average fast-food chicken sandwich. While, at first glance, a fast-food chicken sandwich seems like the very definition of generic, it is, on the contrary, a modern culinary miracle. Even before the actual live chicken leaves the barn in Arkansas or Idaho or eastern Maryland, the wheat for the bun is harvested in Montana. The lettuce is grown in California. The soybeans for the mayonnaise are trucked from fields in Pennsylvania. See what I mean? Even our most commonplace foods have tremendously complex backstories. From originating at the farm (or the hatchery before that) to traveling to the processor, then to the warehouse, then to the restaurant or grocery store or beyond, when it comes to the plain chicken sandwich, complexity reigns.

Everyone must eat. Therefore, one might think that we'd all be invested in understanding how our food system works. But for the vast majority of consumers, as we've emphasized, daily meals revolve around basic considerations of price, taste, and convenience. Objectively, a chicken sandwich might be an amazing feat of modern magic, but it's a trick that consumers have seen so often, the farming wizardry behind the scenes is easily taken for granted.

If you're wondering how this example of large-scale agriculture and the commodity system pertains to you, the new farmer, then reflect back to chapter 5. Remember, no matter how much we wish to stake our own independence, the commodity system will always have an influence on the price of land, on the costs you're able to pass along to customers, and even on your ability to sustainably pay yourself.

Because of this, you must become an educated expert on how our mainstream food system works. This includes being familiar with the different types of markets and distribution channels, as well as the often cryptic mechanics of agricultural economics. As a new farmer, it will be to your benefit to pull back the curtain as much as possible, understanding some heretofore neglected considerations, such as trucking, warehousing, processing, and distribution. Knowing how conventional food is grown will also be useful when customers wonder why prices for your sustainably produced crops are different—this can be a very powerful marketing tool. And, at a minimum, this knowledge will provide a rudimentary framework for how your own systems might one day function, and even give you insights into what you'd rather *not* be doing. It bears repeating that, sometimes, knowing what you don't want to do is as important as knowing what you do want to do.

THE COMMODITY SYSTEM AND THE CHICAGO MERCANTILE EXCHANGE

Most immediately, you need to have a firm grasp of how the Chicago Mercantile Exchange (CME)—the world's largest agricultural marketplace—operates, and why it matters so much to us. On the CME, everything from orange juice to bacon to barley is speculated upon.

Agricultural commodity traders are essentially professional speculators who purchase contracts well in advance of the harvest, otherwise known as futures. These contracts exist only on paper; 99 percent of traders never intend to take physical possession of a tractor-trailer load of corn. Instead, the contract is sold some time in the future (to someone who can actually use it), and the trader either makes, loses, or breaks even on his investment, depending on the real-time price. Professional traders might bid for harvests based on regional weather trends or global consumer demands. For example, they'll bet on wheat if the spring is rainy and farmers can't get into the fields to plant. Or, they'll drive down the price of beef, responding to a new Canadian trade agreement. Between planting, harvesting, and final delivery, the traders endure the risk of fluctuating prices, and big profits or losses occur practically every second of the period in between. This kind of system has been around for millennia, starting when Chinese traders speculated on rice six thousand years ago.

Another reason for learning about the commodity system is because, even though we recommend never relying upon it as your primary sales channel, occasionally, it might be of strategic use. This system offers access to preexisting distribution channels, wholesalers, grain elevators, packing plants, and slaughterhouses, places that receive practically any type of fruit, vegetable, animal, grain, or dairy product. It provides a last-resort safety net if your other marketing plans fall through. It's not an option for every farmer—for instance, small-scale vegetable producers are most likely excluded here. But if you are thinking of growing fruit, grain, dairy, or livestock, then chances are that you'll be able to sell it somewhere on the commodity market.

But understand this: The commodity system is engineered to pay farmers as little as possible for their products. This isn't to say that the market is intentionally malicious, methodically trying to drive farmers out of business. Instead, it's a classic example of Economics 101, the basic laws of supply and demand. When there's scarcity of a product and/or high demand, prices tend to rise. When there's a surplus and/or low demand, prices tend to drop. These numbers can swing wildly from one year to the next, or conversely, they can slowly trickle lower and lower as consumer demand decreases or as efficiencies bring more supply to the market.

Things get even trickier because of federal subsidies, which offer support payments when prices dip. These practices favor large-scale farms that grow the most commonly planted crops (namely, corn, soybeans, wheat, rice, and cotton), and are unavailable, for instance, to small-scale vegetable or livestock producers.

I lived this high-stakes reality personally, and it nearly broke me. In 1996, the year I graduated from college, I took over management of my grandparents' once-successful commodity farm. Assuming that I could operate much like they had for the previous sixty-five years—yet also very aware that my own farming experience was severely limited—I collaborated with an experienced grain farmer to plant several hundreds of acres of corn and soybeans, and sell the crop on the commodity market. Based on soil conditions, rainfall, and total acreage planted, the farmer conservatively estimated we would make a profit of $20,000, or $10,000 apiece. For me, eager to get started—but also facing annual land taxes, overhead costs, and a mailbox perpetually stuffed with overdue, inherited farm bills—the prospect of a $10,000 profit seemed almost too good to be true.

Turns out that it was. Our region was stricken with drought, while the Midwest had a bumper crop of the decade. Suddenly, markets were flooded with grain, so much so that tractor trailers were dumping mountains of corn and soybeans across parking lots in Nebraska, Montana, and South Dakota. There was nowhere else to put it. Naturally, prices plunged.

That $20,000 payday we had planned on? We ended up making $18.16 apiece! I was devastated. But instead of throwing in the towel—as tempting and economically sensible as that might have been—I took a long look in the mirror and considered what I had learned. Ultimately, I realized three things about the commodity market:

- One, although we had made every effort to guarantee the price we would receive, we had practically no control in the end.
- Two, there was only one practical outlet for our product. In other words, we certainly weren't going to sell five tractor-trailer loads of grain through the classified ads or at a roadside stand. In this way, the commodity market essentially had control over us.
- Three, and perhaps most important, I suddenly understood that the main reasons I had wanted to become a farmer in the first place—a sense of independence, creative freedom, and entrepreneurship—were largely false conceits in this system. In the commodity market, I was working for a vast economic machine, one that couldn't care less whether I personally succeeded or failed. As the old saying goes, I was just another widget-maker. At the end of the day, my products were anonymously absorbed into the national food system, and I was paid as little as the market currently offered.

I decided that, from that point on, I would neither try to beat the system nor, as they say, rage against the machine. That would be a waste of my energy. On the contrary, I knew that this market would always be there as a last-resort safety net, just in case my ambitions didn't work out (and nearly twenty-five years later, the commodity system is still going strong). To this day, as a livestock producer, I still keep it as a potential tool in my toolbox, using it once every few years to sell a group of old cows, or when we have an especially large crop of lambs. I also use the prices as fair-market gauges for what replacement stock should cost, as well as for budgeting for feed for my pigs and poultry. But otherwise, it accounts for a fraction of 1 percent of my annual revenues.

Now if this seems like a roundabout way to get to the heart of a chapter titled "To Market, To Market!" then it's for good reason. As the saying goes, "Know the rules well so you can break them properly." Experiencing the mainstream commodity system, both learning and suffering from it, motivated me to pursue alternative paths. And it's through alternative agricultural markets that I have found my version of farming success.

ALTERNATIVE RETAIL MARKETS

We've all heard that nothing worth having comes easily, and branching out as sustainable farmers certainly falls into this category. Agricultural independence, though a desirable and worthy pursuit, comes with its own risks and challenges. How difficult, exactly, is it to strike out on one's own? This is a foundational question for the new farmer, and one we'll attempt to clarify. Although sales of organic food continue to grow by leaps and bounds (see chart below), roughly 96 percent of all food crops still remain destined for commodity-based markets. Hence, the overwhelming majority of farmers rely on commodity outlets for their incomes—and this must be for a reason. Is it because independent marketing is so difficult, or that the commodity system offers full back-end processing, branding, and distribution solutions? Or, sinister as it may sound, is the food system basically rigged, making it next to impossible to operate without its assistance?

These are difficult questions to unravel, and ultimately, they will be up to you to decipher. But the answers will inform your marketing decisions, so you should keep them firmly in the back of your mind. In the meantime, let's explore the vast array of alternative marketing options you have available to you. To market, to market!

Total US Organic Sales and Growth, 2007-2016

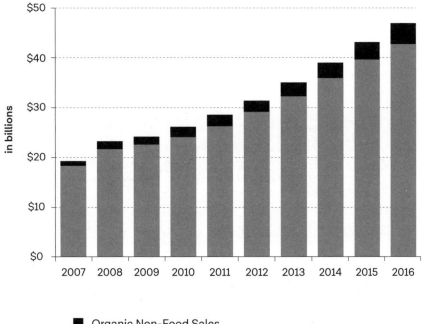

- Organic Non-Food Sales
- Organic Food Sales

Farmers' Markets

So you might be wondering, *What accounts for the 4 percent of alternatively marketed foods?* Great question. With their meteoric rise over the past two decades, the most obvious and enthusiastic response might seem to be farmers' markets (see chart on page 112). They are certainly a component, and they happen to be a major sales channel for me. Farmers' markets are a natural intersection between customers and producers, an economic model that's nearly as old as cities themselves.

This is where I found my footing as a young farmer, waking up early each weekend morning and driving my truck across the Blue Ridge to markets in

National Count of Farmers' Market Directory Listings

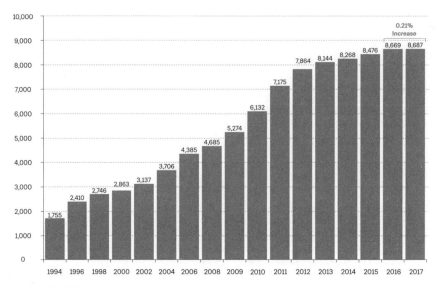

Source: USDA-AMS-Marketing Services Division.
Farmers' Market information is voluntary and self-reported to USDA-AMS-Marketing Services Division.

the Washington, DC, area. It's been rewarding to see their dramatic rise on a national scale, and I hope the day arrives when every small town has a bustling, producer-driven market at the core of its community.

However, while farmers' markets have been the poster child for twenty-first-century agriculture, the fact is that only a fraction of that remaining 4 percent is represented by producers who make their living at these markets. Out of more than two million American farmers, the number of those who depend on farmers' markets for their revenue is probably far smaller than we'd like to imagine.

From the USDA, we know that there are roughly nine thousand farmers' markets in the country. With an average of ten full-time producers per market, this translates to only sixty thousand—that's far less than 4 percent of the total number of American farmers. Then, remembering that the average market only operates for six months, as well as the fact that some producers attend multiple markets, the number of farmers would surely plummet even further.

In short, despite their popularity, farmers' markets aren't the sales panacea that we might imagine.

If these markets don't comprise the primary sales channel for sustainable growers, then where are these producers selling their goods? It's difficult to pin down, exactly, but the answer is probably a little bit of everywhere.

A good analogy here is renewable energy. Coal, for many decades the primary source of our electricity, has been increasingly replaced with solar, nuclear, natural gas, wind, hydro, and biomass energy—yet with no single winner-take-all technology. In a similar fashion, pioneering farmers who seek an alternative to the commodity system have had to be both innovative and opportunistic, utilizing a broad portfolio of outlets—often simultaneously. Besides farmers' markets, below is a list of the most common ways non-commodity farmers market their products, nearly all of which we have utilized at some point over the years.

CSAs

This is an acronym for *community supported agriculture.* In a CSA structure, individual customers pay months in advance for a "share" of the farm's produce, which is then delivered or available for weekly pickup once the growing season arrives. "Produce" is typically defined as vegetables, but can also be meat, dairy products, fruit, grain, honey, or some combination thereof. Some CSAs supplement their primary offerings with à la carte additions from other local farms; for example, a gallon of milk and a dozen eggs alongside a seasonal vegetable allocation.

CSAs offer a compelling opportunity for farmers and consumers alike. The farmer receives an advance cash investment ahead of the growing season (hence, the "supported" component of CSA), which is especially vital during a time of year when revenue is typically scarce. The consumer, meanwhile, is assured the freshest, in-season produce throughout the spring, summer, and autumn.

Roadside Stands and On-Farm Stores

From hyperseasonal strawberry, watermelon, or pumpkin patch pop-ups, to permanent, year-round locations, practically everyone has seen (and hopefully enjoyed) a roadside stand or on-farm store. While they typically feature produce that's in season, many stands remain open year-round by diversifying with additional agricultural products (including jams, honey, pies, craft beer, cider, or cheese), providing a wide spectrum of offerings. Of course, the flip side of this would be the sweet-corn-laden pickup truck alongside the highway, perhaps only selling for a week or two of the entire year. This is how I supplemented my summer income as a fourteen-year-old, illegally driving our "farm use" Ford across the highway and selling fifty dozen ears in a truck-stop parking lot.

Pick-Your-Own

Getting folks to come out to the farm and soak up the ambience as they harvest their own apples, blueberries, or even Christmas trees is another great way to retail. It's important to note that, generally speaking, drawing customers to your farm is the most efficient way to market your food or products. There's no transportation cost, and you've potentially eliminated storage and warehousing. Adding to this, if you can provide a fun experience where your customers do their own harvesting, then subsequently pay you for it, the numbers can be very favorable!

Naturally, a pick-your-own operation comes with its own set of challenges. This includes limited seasonality, extra marketing, insurance against customer injuries, and additional infrastructure such as restrooms, shelters, and extra staff. Adding to the potential cost, even a few weekends of bad weather can really dampen your returns. But for a farmer who doesn't want to leave the property, or who prefers a short window of production, a pick-your-own operation can be just the right fit.

Direct Shipping

The creation of the refrigerated boxcar in the late 1800s revolutionized our food system, allowing perishable vegetables and meat to be safely transported long distances. For nearly a century afterward, though, this technology was primarily available only to large-scale wholesalers and meat-packers, who either owned the boxcars outright or subcontracted them via the railroad lines.

However, the second half of the twentieth century ushered in a new era: overnight shipping and air freight, available at affordable prices for individual consumers. Companies such as FedEx, UPS, and even the US Postal Service suddenly made it safe and economically viable for a small fisherman to ship Cape Cod lobsters to Los Angeles, or an orchard in Los Angeles to ship citrus to Cape Cod. While the impact of an increased carbon footprint is worthy of consideration, this sales channel is especially useful for farms that might be otherwise isolated by geography or unable to access more than a small local population.

On-Farm Restaurants and Agritourism

Increasingly, customers are seeking unique, seasonal culinary experiences that directly connect them to where their food originated. But instead of pulling into a restaurant parking lot, in this instance, they are driving out to the actual farm, enjoying a meal beside the fields, pastures, and barns where their food was raised. Intrepid farmers are either collaborating with local chefs to create an occasional on-farm restaurant fusion or opening a fixed location restaurant on their own properties.

The portmanteau "agritourism" is just as it sounds: It emphasizes agriculture as a tourist destination. City-dwellers have been escaping to the country as far back as the early 1800s, and this American tradition has changed little

Reason for Trip	Number of Responses (millions)	Percent of Respondents
Enjoy rural scenery	53	86
Learning where food comes from	44	71
Visit family or friends	40	63
Watch/participate in farm activities	41	64
Purchase agricultural products	27	39
Pick fruit or produce	27	43
To hunt and fish	16	27
Spend a night	19	8

USDA Forest Service

over the intervening centuries. According to a USDA Forest Service survey (see chart on page 115), there's a wide spectrum of reasons as to why people seek out the countryside, in general, and farms, in particular.

As for on-farm events themselves, some are special once-a-year occasions (like an annual Farm Day or Harvest Festival), or sometimes revolve around autumn weekends (like corn mazes and cider making), while other events integrate a permanent, on-premises restaurant experience as a component of daily farm life.

WHOLESALE MARKETS

Wholesaling is when an intermediary purchases produce from a farmer, subsequently finds a market for it, and resells it again at a higher price. There are many reasons why a farmer might choose to sell produce to wholesalers: a high volume of product, a narrow seasonality, or the desire to avoid the expense of warehousing, distribution, and marketing. As the local and sustainable food movement has gained momentum, using wholesaling channels has become a more viable financial option for farmers, because the gap between pure retail and traditional wholesale pricing (which is often 50 percent or more discounted from retail) has narrowed to as little as 20 percent in many instances. Below is a list of the most common wholesale markets that value fresh produce.

Fresh Produce Wholesaler

Maybe you'll end up growing a lot of one crop, or perhaps yields will exceed your expectations, and you'll need to move a lot of produce at once. This is where a wholesale vegetable or fruit market can be extremely useful. Wholesale markets operate much like commodity markets, but instead of receiving cattle, corn, or cotton, they take in fresh vegetables and fruits (alternatively, these can also be sold to packing houses, but this is typically on a much larger scale). Wholesale markets are most often near major cities, where perishable produce can quickly be bid on, distributed, and offered for sale to restaurants, groceries, and other fresh-food markets.

Sometimes, this can be an auction, where the farmer receives the price of the highest bid, while other times, producers are able to lock into guaranteed contracts in advance with specific buyers. In either case, the prices are far below retail value, since the wholesaler must make a profit by brokering produce to other parties. Many farmers actually rely on the wholesale market as part of their distribution plans, because their production efficiencies—and lack of other marketing costs—allow them to make a profit even at these reduced prices. From our experience, the wholesale market can be a useful tool in our sales channel toolbox.

Food Hubs

Similar in structure to wholesale markets, food hubs are regional produce aggregators that offer a wholesale price to farmers, taking over the storage, distribution, and sales of large amounts of produce. Where they differ, specifically, is in their mission. Food hubs intentionally source from local and sustainable farmers, and distribute directly to customers who place a higher value on where their food comes from. Hence, these clients are typically willing to pay more than a traditional wholesale price.

These are a recent addition to the food distribution landscape (most have come online since 2008). Many food hubs also intentionally focus on small- to medium-size farms, offering an outlet to producers who might not provide enough volume for a traditional wholesaler. Food hubs have been especially active in bridging gaps between farmers and institutional buyers, for instance, providing fresh, local food to hospitals and schools.

Farm-to-Table Restaurants

Since the early 2000s, farm-to-table restaurants have been gaining momentum and popping up all over the country, and presently they are enjoying an unprecedented surge in popularity. Whereas typical restaurants source ingredients from major wholesalers, offering dishes made of comparatively generic produce, farm-to-table restaurants feature menus that work directly in tandem with regional farmers, providing local seasonality alongside traditional, year-round fare. With such a multitude of chefs interested in this theme, as well as robust customer response, it's a bright opportunity for producers looking for steady, year-round income.

Grocery Stores

Although the overwhelming majority of the food found at grocery stores comes from large wholesale distributors and packing plants—and a significant portion from international sources—local farmers have long maintained personal sales relationships with nearby stores and regional chains. When I was a teenager, as part of a crew, I picked sweet corn bound for grocery stores in suburban Maryland; we filled an entire school bus, stripped of its seats. Many years later, I was recounting this story to a regional Whole Foods manager, who informed me that they now court farmers of all production sizes, not just those who could fill a bus.

In short, grocery stores can be a viable wholesale option for small- to medium-size farmers, but success will vary from region to region. If you are looking to collaborate with your local grocery stores, simply schedule a meeting with the local manager (of course, many months in advance of planting), bring a production chart, and start a conversation. It can be just that simple. I know dozens of farmers who have done precisely this, and are now fixtures in their local supermarkets.

Cooperatives

Agricultural cooperatives have a storied American history, with farmers banding together since colonial times to market their products. In any co-op, members pool their resources for greater purchasing leverage—thus, typically creating lower costs for goods or services. In an agricultural co-op, however, farmers not only gain these benefits but also establish a steady product stream with multiple like-minded members, as well as create a brand with a national reach. Think Land O'Lakes, Ocean Spray, and Sunkist. Organic Valley, based in Wisconsin, is the largest organic cooperative, made up of two thousand members.

Depending on your crop, there may be a good opportunity to sell to a local cooperative. For fresh produce growers, a good example is Tuscarora Organic Growers Cooperative in central Pennsylvania. Here, the mechanisms of sales and transport are just like those in a food hub, but co-op members have the additional benefit of direct input and voting to impact future business decisions.

WHICH IS RIGHT FOR YOU?

Clearly, the options for marketing produce beyond the mainstream commodity system are myriad, limited only by the creativity needed to conjure them into reality. Accentuating this list, we could include food trucks, buying clubs, distilleries and wineries, sporting events concessions, catering companies, and home-delivery services. It's a list that's practically endless. Suffice it to say, the world is a hungry place. If you can package it, cook it, bake it, or somehow put it on a stick, chances are that someone will buy it.

Though many of these outlets might appear familiar at first glance, each is fraught with its own pitfalls and challenges, begging multiple questions: How do you choose a specific market channel? Should you attempt to supply multiple sales outlets at once? How do you make initial connections and gain a foothold? And perhaps most important, knowing how difficult farming is, should you attempt to be both a producer and a salesperson? Lie awake restlessly for several nights trying to answer these questions, and you might understand why commodity farming—with all the answers already built into the system—remains the overwhelmingly popular choice of most producers.

Let's start with being strategic. *Take what's free to you, and maximize it.* If there's a farmers' market in your county or town, visit it. And not just once. Go there weekly for at least two straight months—even longer, if possible. See what's available for sale. Shop. Just as important, observe what's actually selling, and in what quantities. Try to remove your preconceived notions about

TAKING THAT FIRST STEP

When Allen and Heather McWilliam started Tannadice Farms on Vancouver Island, British Columbia, in 1974, they began as a commercial hog operation. Fluctuating commodity prices were always a challenge, and after a few years of enduring rock-bottom prices, they knew they had to radically shift their strategy, or the farm would fail. With no marketing experience whatsoever, Allen began calling local grocery stores and offering custom cuts of pork. The couple slowly grew and built upon these connections, and, within several years, they were able to sell all of their products this way, no longer reliant on the commodity system for their livelihood.

what's popular and what's not. Look for a scarcity or complete absence of a certain type of product. Conversely, look for products that sell well regardless of how many other farmers are offering it. Pay attention to anomalies. If the traffic at market plummets on a rainy day, or when temperatures soar above 100°F, then extrapolate this information. Will there be enough customers for you to make your sales goal each time?

Consider the location. Whether the market is near a highly populated neighborhood or far from town and inaccessible to foot traffic can make a huge difference. Maybe there's a grocery store nearby that might compete for customers. Or, thinking laterally, that same grocery store might be an asset, supplementing offerings that the farmers' market doesn't provide. Dig deeper, and learn where that grocery store sources its fresh produce.

While you're at it, consider your favorite restaurants. Do any offer a farm-to-table theme or component? Ask the non-farm-to-table restaurants in your area if they'd be willing to offer local products. Research your personal connections to the chef, or to your local produce manager, as well as your other connections in your community. These are the folks who might be interested in buying your produce. This free information can be accumulated and pondered throughout the course of a normal day, with little or no additional effort on your part. Easy, right?

CUSTOMER CONNECTIONS

You know more people than you think you do. Best of all, those people know other people. Consider for a moment how social media works. It's amazing to see how many of your friends know one another, even though you had no idea they were connected beyond your individual friendships.

Now think about how this plays out in person, meeting a stranger for the first time. Are you more likely to trust someone if they make an unsolicited sales pitch to you out of the blue, or if they are first introduced to you by a mutual friend? Like any good farm boy, my grandmother raised me to be exceedingly polite. I'll freely admit, however, that I'm more likely to listen to someone who has been introduced to me by a trusted friend.

Utilize your connections—respectfully, of course. But don't underestimate them. What if you don't have an in when it comes to a certain opportunity you'd like to pursue? Even if you have no preexisting connections, nothing speaks as clearly as already being a customer when it comes to broaching a business deal—and by this, of course, I mean supporting your fellow business-person by purchasing their products or services.

Imagine it. Would a farmer be more inclined to take time out of their busy schedule to offer advice if you've been shopping with them steadily for months or if you are a complete stranger who shows up with a laundry list of questions? I can't begin to tally the number of emails and phone calls I receive from people I've never met, asking for "just a few minutes" of my time to bounce farming ideas off of me. Almost invariably, they have questions that would require hours of explanation to answer appropriately. In short, treat others with the consideration that you'd wish for yourself, and more doors will open to you.

Hence, in order to successfully cultivate a profitable, sustainable alternative market channel—or in all likelihood, multiple alternative market channels, you'll need to follow three main steps:

1. Identify a need.
2. Find the angle that suits your local customer base as well as your personal strengths.
3. Foster your relationships.

Pursue each of these steps over the course of months—preferably a full year—before making up your mind about what you'll ultimately decide to grow, or how you'll market it. This will be time very well spent.

As a farmer, it takes courage to break away from the mainstream commodity system. But, just as important, it requires planning, perspective, and insight. These aren't acquired overnight, or garnered from reading a book. Instead, real-world application is needed. Fortunately, this opportunity is yours alone for the taking. No one else has your exact combination of passion, connections, and unique capabilities.

In the next chapter, you'll find strategies for how to best utilize these alternative marketing channels, carefully examining them from an insider's perspective.

CHAPTER REVIEW QUESTIONS

1. Have you ever subscribed to a CSA? If so, what could have been better about your experience, and how would you improve it? If not, ask friends about their experiences (on social media, if need be), and follow up with the same question.

2. Where is the closest vegetable wholesaler to you? The closest livestock auction? The closest farm-to-table restaurant? The closest farmers' market? Spend a few hours over the next month visiting each of these, and observing.

3. List five crops that no one in your area is growing. Discounting specific crops that simply won't thrive in your climate, why do you think this is the case? Which of these crops would best suit a local need, if you started growing it?

CHAPTER EIGHT

TO MARKET, TO MARKET!
PART TWO

Forrest

There are few things I've found more challenging—or ultimately more rewarding—than figuring out how to effectively sell my products. After all, I spent my college years studying rocks and Shakespeare, not marketing, sales, or business. Aside from helping my grandmother sell eggs to her fellow churchgoers, my exposure to marketing was limited to reading *Death of a Salesman* and watching the comedy *Tommy Boy*. This isn't a strategy I'd recommend to anyone!

When I returned to the farm after college, growing food took up practically every second of my time. But that wasn't the end of my responsibilities. Not only did I have to overhaul our aging commodity-based production systems, but I had to simultaneously figure out how to get my products into new sales channels, manage my ever-shifting inventory, and, all the while, balance seasonality into the equation. Those first couple of years, boy did I make a lot of mistakes. I pursued restaurants, catering companies, wholesalers, grocery stores, farmers' markets, and home-delivery services. Before long, I felt like a dog chasing its own tail.

It took a huge amount of time and effort, but, gradually, I found my own mix of ways to market my products. No doubt, it took far longer than I would have preferred. Looking back, though, I'm not sure that I could have done it any faster, better, or differently. Local markets and demographics, limited production capabilities early on, and my personal inexperience all combined to make my sales channels uniquely challenging.

Due to these factors, when it comes to salesmanship, it's next to impossible to mimic someone else's strategies exactly. In fact, I continue to tweak and modify my own marketing strategies to this day. Presently, I sell about 85 percent of my products through direct marketing (this means selling straight to the customer—in my case, either at farmers' markets or in my on-farm store), and the remaining 15 percent via wholesaling (in my case, to restaurants, other stores, and home-delivery services).

I also adapt my product line each year, based on consumer demands. Sometimes I introduce new products (for example, beef jerky, salamis, or pet snacks), sometimes I expand distribution (perhaps an additional farmers' market location), and sometimes I even shut something down that once worked in the past. There's no silver bullet, and there's no ironclad method. The target is always to make a profitable sale, but that target is constantly moving, and it's rare if I don't use more than one arrow (marketing strategy) to hit the bull's-eye.

However, if there's one general, overarching rule about markets, one that you can immediately apply to your own operation, it's this: What you grow will have an enormous influence on how you can sell it.

THE FOUR MAJOR FACTORS OF PRODUCTION

You've chosen your farm site, your systems, and now you're all set to start producing vegetables, turkeys, apples, or milk. So far, so good! But exactly *what* you ultimately decide to grow—when it's in season, how difficult it is to harvest or process, how perishable it is, and how challenging it is to transport—will greatly impact your ability to find a suitable outlet, and get you paid for your hard work.

Always keep in mind that unless you're directly marketing all of your products straight to the consumer, the considerations listed above aren't just your problems. They also become the problems of any market that is about to receive them. For restaurants, groceries, or wholesalers, perishability (waste) and storage costs must all be factored in and passed along to the final customer. Seasonality affects everyone.

Imagine having no watermelons for eleven months, then suddenly finding yourself with a tractor-trailer load that needs to be moved in a week or two!

On the other hand, with thousands of customers, it doesn't make sense for a grocery store to receive only a dozen melons, either. The sweet spot is somewhere in the middle, but there's no way to know the precise number. Even so, these are the kinds of details that must be negotiated months or even a full year in advance; otherwise, that's a lot of watermelon seeds you'll be spitting come harvest time.

What you grow, *when* you grow it, *how much* of it you grow, and *how long it will stay fresh* are the four major factors that will forever impact your ability to sell your products. The more you can broaden or limit these variables—for example, extending your offerings year-round, or finding ways to reduce the transportation of your most perishable products—the greater your chances will be of finding the right markets to fit your farm.

IDENTIFY YOUR MARKETS

Of course, alongside all the challenges, everything you grow will come with its unique advantages, as well. On my friends Don and Delores Magnani's farm, located an hour south of Washington, DC, they grow fresh figs, a sticky fruit that's too delicate to ship long distances. But because they take the extra time to carefully pick them the night before they are to be sold, then truck them straight to the farmers' market themselves, they make an absolute killing when the fruit is in season. Their stand is easy to recognize; it's the one where people are lined up twenty deep! They've effectively cornered the market on this fragile fruit.

My friend Mark Toigo in Pennsylvania grows tomatoes in greenhouses throughout the winter, and sells them to regional Whole Foods locations. But much like the figs, these heirloom tomatoes have delicate skins, and can't take much jostling on tractor trailers and loading docks—his tomatoes are distributed from New England down to North Carolina. To solve this problem, the tomatoes are placed in recyclable clamshell containers, six to a pack. They're not only protected but also they carry his name and logo as well.

Providing a product that's typically only available via the commodity system, my friend Steve Ernst in central Maryland has diversified his farm of many hundreds of acres into non-GMO grains, offering blended animal feed to local farmers and hobby livestock owners. Steve and his sons have constructed their own storage bins and bagging operation, eliminating their former reliance on trucking it to the closest rail yard. These days, people come to him with their orders, arriving in SUVs and pickup trucks.

So what do these three very different producers have in common? They've each identified both the advantages and shortcomings of their individual products, accentuated the positive, and honed in on a marketing strategy that best suits their production.

As stressed in the previous chapter, the sustainable farmer should avoid the commodity market and its fixed pricing structure as much as possible. This, of course, requires a greater amount of work. Though you're positioning yourself for a potentially higher payday, you're also introducing all sorts of additional risk and extra effort into your life. This doesn't mean that you should be afraid to dip your toes into this water. As we'll repeat over and over, risk and extra effort are always where the most money can be made. But this is why you'll need to invest lots of time brainstorming and identifying all the markets that are available to you, and attempt to figure out which market—or more likely, which combination of markets—will best accommodate your needs.

Depending on where your farm is located, you might have a plethora of marketing options, or, at first appraisal, few whatsoever. Whatever your personal circumstance may be, always remember this: You have many more options than you realize at first glance. Yes, location certainly matters. But so does perspicacity—the art of perception combined with insight. Once you really concentrate on identifying marketing opportunities, you may be surprised by the ocean of possibilities that come flooding your way. Sometimes, it can even be hard to shut off the flow of ideas.

Anyone can identify the obvious candidates—grocery stores, farmers' markets, and restaurants. No great challenge there. But if these are the most likely choices, then rest assured that others are thinking this way, too. It's far more useful to sit down and draw up a list, identifying every single retail and wholesale outlet that comes to mind. Keep growing this list for months, as you come up with new ideas. Thoughts will generate more thoughts, unlocking opportunities that you hadn't previously considered.

Ideas can be big or small, depending on your scale and what you grow. Your local bakery might need twenty pounds of herbs each week, or there's that coffee shop that wants to feature locally sourced food on their menu. The bed-and-breakfast in town needs fresh produce and flowers each morning, and the gym with five hundred members would be a perfect drop-off spot for a paleo-oriented CSA.

Perhaps these opportunities are too small for the amount of food you're producing. What about approaching a regional hotel chain about supplying their in-house conference needs? Or nearby wineries and brewpubs about becoming a featured supplier? You could purchase your own two-door refrigerators, brand them with your logo, and place them in country stores in your area. Better yet, open your own country store, and not only sell your fresh produce, but process them into value-added products (e.g., snack sticks, jams, and baked goods).

These are just a few ideas straight off the top of my head, typing in stream of consciousness without pausing. It was an easy list to write because I've personally tried nearly each one of these marketing outlets myself! You might look at these suggestions and dismiss all of them; or, perhaps one or two opened a door in your mind. And just to be clear, there's nothing inherently wrong with approaching the more obvious groceries, farmers' market, or restaurants—I sell to each of these myself.

Rest assured, there are dozens upon dozens of ways to market your products. As an old farmer once told me, "I've seen all sorts of trends come and go over the years, but one thing always stays the same: People gotta eat!" Find where people eat, buy food, or go to be entertained—or, ironically, to exercise *off* the food they eat—and you'll find customers. Again, one of the best ways to do this is to identify a need in your community, or find a certain angle, especially one that protects you from copycats. Opening my on-farm store checked each of these boxes for me.

It's always tempting to think that location is the biggest limiting factor when it comes to marketing. I can hear some of you now, "Well, marketing is easy for Forrest. He lives an hour from Washington, DC. I live in Oklahoma." But as you'll learn in the next chapter, you must train yourself to allow this conventional (and often self-defeating) thinking to flow past, in order to discover your own productive path. Yes, fate has it that I live near a sizable metro area—with a population of six million, tied with Atlanta for ninth largest in the country (for perspective, Tulsa and Oklahoma City are *each* home to roughly one million potential customers). But I also happen to live in one of the more fertile agricultural regions of the entire country, which means there are hundreds upon hundreds of other farmers offering very similar products all around me.

True, perhaps you're much more isolated, but that's when you identify a need, and maximize it. There's simply no easy recipe for any of us. At the end of the day, it's your duty to adapt, transforming where you live into your unique advantage instead of a liability. You can be one of the most talented growers in the world, but not make a penny of profit if you can't solve this riddle.

Don't forget, this is why the commodity system exists, and is so prevalent—it takes what farmers grow off their hands, and eliminates the need for marketing and logistical problem-solving. But the compensation the farmer receives is commensurate with this service; the extra effort of marketing and distribution is done on the farmer's behalf. Want to make more money? Then marketing and distribution issues are up to *you*.

LOCATION, LOCATION?

There are, of course, always extreme exceptions to the location conundrum—farms located so very far away from customers that there's no practical way to access local markets. For these producers, perhaps a regional co-op or a wholesale distributor might be the best marketing path, combined with mail-order shipping of value-added products to promote the farm's brand. The internet and social media have radically changed the playing field when it comes to selling your products, and you need to be savvy. In short, don't convince yourself that you have no alternative to commodity markets. While farming in a remote location will certainly impact your choices, never rely exclusively on the commodity system to solve this problem for you—it won't.

PRIORITIZE YOUR MARKETS

So now you've made a very thorough and creative list, and are all geared up to go knocking on doors, make cold calls, or fill out vendor paperwork. Not so fast, farmer. Now's the time to prioritize your list, factoring in distance, transportation, delivery scheduling, volume, and overhead costs. In other words, you need to rank each of your options with pragmatic considerations as to how they best fit into your system. Making the right choices here is often the difference between sleeping well at night and waking up in a cold sweat.

Sometimes, you might identify a big opportunity that seems almost perfect. Early in my career, I nearly had a deal with a local university cafeteria. They wanted to purchase five hundred pounds of hamburger a week—the equivalent of five cattle each month. This would have been a substantial order for me, and I tried very hard to make it work. We went back and forth for several months, trying to work out the logistics. In the end, though, the absolute lowest price I could offer was $4.00 a pound, and the highest they could offer was $3.75.

The math was black-and-white to me, but the purchasing agent was flabbergasted that I wouldn't lower my price twenty-five more cents. They were accustomed to dealing with wholesale meat from the commodity market, not a farmer who was trying to account for his full cost of production. Between operations costs, processing, storage, and delivery, there was simply no way I could sell my product at less than $4.00—I might as well have sold my cattle to the livestock commodity market at that price. In the end, even though I would have grossed nearly $100,000 a year, I wouldn't have made a penny of profit for all the extra work and expenses.

Not only did this arrangement fall through, but I had been so focused on this deal that I had never taken the time to create a list of additional sales channels—a marketing plan B. I soon found myself scrambling for revenue, delivering roasts to restaurants in Columbia, Maryland, then turning around and trucking T-bone steaks to a catering company in Middleburg, Virginia. Then, I dashed off to a tiny health food store in the Shenandoah Valley, stocking a freezer that needed a defrost back in the Eisenhower era.

A wise farmer once told me, "You've got to kiss a lot of frogs before you find a prince." Sure, I was sticking to my prices, but I was kissing an awful lot of frogs along the way. Worse yet, I was running myself ragged doing so. I eventually had the good sense to return to my list of marketing priorities and focus on the avenues that provided the clearest path to profit—outlets that offered synergies with my production schedule. After several years of trial and error, my list eventually looked like this:

- Reduce my delivery schedule to only weekends, ensuring steady, on-farm production at least five consecutive days.
- Limit deliveries only to locations more or less within a straight line of where I was already going (no individual outliers).
- Only take on new clients who place at least $500 per order, or a new farmers' market that would generate at least $1,000 in daily revenue.
- Figure out a way to bring more customers out to the farm, reducing the need for deliveries and other marketing overhead.

In the end, I realized that to move all of my product reliably—as well as secure a price that worked for our farm—I had to develop a multifaceted strategy. This included retail sales, strategic wholesale deliveries, and all the while, encouraging folks to come out to the farm to shop. In other words, instead of relying on one big client, or running around to dozens of small locations, I needed to sensibly diversify.

AN ORGANIC FARMER IN COMMODITY LAND

How does sales diversification work if you grow a large-scale commodity product? Paul's Grains in Laurel, Iowa, is a perfect example of this. Steve Paul intentionally plants many varieties of organic, heirloom grains (wheat, barley, oats, corn, buckwheat, rye) to hedge his bets should a saturated commodity market drive prices lower. He keeps strong relationships with multiple distributors, sending certain grains in one direction, if the demand is particularly strong. He also maintains an on-farm store, where his family mills flours for sale to his local community.

DIVERSIFY YOUR MARKETS

We should always have diversity in our sales channels, and as sustainable farmers, we take this cue from nature. Just as a monoculture is a man-made construct, ever in conflict with the reigning ecology, so, too, is reliance on only one market to support our farms. The parallel here is uncanny. *Mono-* is typically fragile; *multi-* is usually strong. Nature insists on balance and resiliency, and diversity provides this, even when it comes to marketing.

Having several sales channels at once teaches you to be efficient, reveals unexpected opportunities, and sheds light on the strengths and weaknesses of your systems—as well as your personal temperament. It's nothing short of economic biomimicry. When properly balanced, market diversification makes your farms and your balance sheets stronger.

This is what I continue to do to this day, receiving feedback from multiple markets in real time. Some weekends, farmers' markets are mysteriously slow, while my on-farm store has a record day. Other times, the restaurant calls and orders double their usual amount, while the home-delivery service takes a pass for two weeks straight. Or, the next week, all vice versa. Having diverse sales channels not only provides multiple streams of revenue (keeping all-important cash flow reliably rolling) but also it hedges your bets if one particular market stumbles. Maintaining multiple markets allows you the flexibility to adapt, make corrections, or ride out short-term sales turbulence. And take our word for it, waking up on a Monday and facing an entire workweek of chores, production, perspiration, and payroll, this economic of peace of mind is something close to priceless.

So what's the optimal mix of diversification? The short answer is that this can only be answered by you—and over a long period of time. But take the initiative to establish multiple channels, even if your sales seem to be humming along perfectly. Markets can change practically overnight. The trick is to find the right blend of options that works for your individual farm, and keep tweaking it toward maximum profitability.

APPROACH YOUR MARKETS

In chapter 2, we spent a lot of time talking about personal temperaments, and let's be honest here: Farmers, in general, are not extroverted people. A life in agriculture typically requires long stretches of solitude, silence, and contemplation. This isn't exactly the résumé of an outgoing salesperson—more like a librarian.

So is it reasonable to expect someone who hardly speaks to anyone other than his sheep to suddenly become a charismatic sweet-talker, cold-calling distracted chefs and drumming up business for his farm? Of course not. When I first started out, my anxiety was so strong that I'd have sooner poked a wasp nest with a stick than walk into a store and deliver a sales pitch. At least I could outrun the wasps! Running out of a store, on the other hand, makes for a pretty awkward first impression.

Your personal temperament will play a huge role here, and you should be aware of this. While you might be the most passionate and knowledgeable advocate for your farm, sometimes that energy is best left precisely where it already flourishes—in the hard work of growing fresh, abundant food. But when it comes to sales, are you the best person to pick up the phone, knock on a door, or send an email? Perhaps you are . . . but maybe not. Or maybe you only need a little self-confidence, tutelage, or practice to build your skills. It's a very personal question to face, but one that is incredibly important to answer for the success of your operation.

You can be a superior producer, an excellent bookkeeper, or an all-star salesperson, but try to be all three of these at once, and something will have to give; there's only so much time in the day. Yes, you can probably squeak by for a couple of years doing it all—and you might even have to during the early stages. Eventually, however, in order to keep both you and your farm sustainably balanced, you're going to need some help. Production requires lots of experience. Bookkeeping demands excellent math skills and attention to detail. As for sales? It requires knowledge of the product, lots of enthusiasm, and perseverance. If you have to outsource any one of these important tasks, sales will usually be the safest bet.

This is typically where you might hear someone say, "Just ask your partner," or "Ask a family member," or even "Ask dear old Mom" to become your salesperson. But not all of us have a partner or a family nearby, or even a sweet old mom willing to march off to sell ten thousand pints of strawberries. Besides, who really wants to put their family or partner under such pressure? They already have lives of their own.

Instead, it's a far better strategy to temporarily hire a professional salesperson to work on your farm's behalf. As the old saying goes, "Everything is sales." The world is practically overrun with salespeople looking for their next gigs. Identify a salesperson with whom you feel comfortable, get them educated and excited about your products, then pay them either a short-term hourly salary or on commission. Then, when they have identified and organized enough sales channels for your specific needs, pay them a retainer to be available for the next time you need their services. Maybe you'll expand production, or maybe a market won't pan out. Far better to pick up the phone and call your part-time sales representative than to suddenly be left high and dry.

What if, however, you *are* a good marketer, someone who can't wait to get out there and tell the world about your amazing products? That's great! The world needs more enthusiastic promotors of fresh farm food. But, again, be judicious in the use of your time and energy. It's very hard to be an expert at multiple things, and farming already requires high levels of diversified expertise, not to mention endless hours putting it all into action. Balance is the key.

When I first started out, I had the benefit of several family members who volunteered to help with sales (my government-employee father actually wheeled coolers of frozen chickens onto the commuter train to Washington, DC, each week, and sold broilers cubicle to cubicle). No doubt, this was extremely helpful. It was still up to me, however, to identify and approach long-term sales channels. This entailed applying to farmers' markets, walking into stores and speaking to the managers, negotiating deals with chefs, and following up on endless emails. No doubt about it, it was exhausting, and, until I got all my sales outlets secure, my production and bookkeeping surely suffered as a consequence.

By trying to approach all those markets myself—instead of focusing strictly on production, and hiring someone to help me with sales—I easily set myself back five years when it came to syncing up my markets with my production. Five years is a lot of lost time, and with hindsight, hiring a temporary salesperson would have likely paid me back tens of thousands of dollars. Again, you can possibly do it alone, and certainly you might even have to for a little while. But knowing your personal temperament and allowing others to help you will go a long way toward successfully approaching and securing your markets.

BRAND YOUR FARM FOR MARKET

I can't overstate the importance of a memorable brand in today's culture. It might seem shallow, and I certainly wish it weren't the case, but consumers associate logos and brand names with their favorite products. Despite my inclinations to rebel against this, there's a reason I own three of the same pickup trucks and bought a particular tractor. Like other consumers, I associate a high level of quality with each brand. Your products should generate the same response.

We'd be naive not to take away some lessons here from the big guys. Regardless of our personal feelings about branding, there's no denying that it works. There's something deeply tribal about names, images, and even colors being tied together with a story. Having a memorable farm name with accompanying signage, labeling, and a social media presence is no longer some fancy luxury. On the contrary, these can be extremely easy to create at very low cost. Free design tools exist to help you make logos and design websites, and branded labels are less than five cents per sticker. A handsome sign at the end of your lane will cost you a few hundred dollars, yet generate a decade's worth of otherwise free advertising. And, after setting up your website pages, maintaining a social media presence should account for little more than five minutes of your time here and there.

Megacorporations attempt to co-opt farming imagery on a daily basis, using wholesome scenery and rural vistas to promote their products. Why shouldn't we snatch it right back from them? We have what they can only dream of: authenticity. Independent, sustainable farmers can take a page from this playbook, and push back with the same methods. So for heaven's sake, at the very minimum, put your name and logo on the side of your truck. Each time you drive down the highway, you're making hundreds, if not thousands, of people aware of your farm.

Simple branding efforts add up. Can people remember the name of your farm, your website, and your logo? The answer to this question might be the difference between an enthusiastic new customer and someone who lives a few miles away but never even realized your farm was there.

FINE-TUNE YOUR MARKETS

You finally have a great mix of markets, a recognizable brand, and someone to help you with sales, if you decide to grow your business further—or help if something goes off the rails. Wow, you're off to an amazing start. Now all that's left to do is to focus on production, load up the truck, and sit back each weekend while the money comes pouring in. You're finally living the dream.

As you surely know by now, farming is *never* that easy. Neither is marketing.

Even when things are really rolling along, you must always remain vigilant, assessing real-time information, and doing your best to increase efficiencies to grow your profits. In order to do keep your sales really humming, we must implement the following strategies.

Offer Great Customer Service

The hardest thing in sales isn't necessarily getting a customer; it's *keeping* a customer. You've put all this effort into finding a market that's a perfect fit for your farm. Do you think the hard work ends when you make a delivery, and point your truck back to the ranch? Of course not. At a restaurant or a store, get to know the staff, the manager, and the owner. Ask them what else they might need, what you can do to help, or what you could be doing better. At a farmers' market or retail outlet, learn your customer's names and their likes and preferences. Provide cooking tips. At the warehouse, bring the guys some doughnuts and coffee, and stick around a few minutes to shoot the breeze. You take care of your customers, and, chances are, they'll take care of you, too.

It's a far better use of energy to keep a customer happy than it is to convince a new customer to purchase your products. Best of all, when a customer really, really loves your product, they will start telling their friends about you. This is the best advertising of all.

Make Things Easy

Many years ago, I was riding a train cross-country and struck up a conversation with a retired chef from Philadelphia. After I told him I was a farmer and that I sold to restaurants, his eyes lit up.

"You got a good thing going," he said. "Now lemme give you some advice." He leaned forward, ticking a list off blunt fingertips. "If you put out a good product, and you always deliver when you say you will, and you don't run out. . . ." He clapped his hands together. "Bingo! You're a piece of gold to a chef. *Gold.* You'll have business for the rest of your life."

In a nutshell, he was telling me to make things easy for the customer. I took his advice to heart. Whether it's working with a restaurant or selling at a retail store, or even arranging a display at the farmers' market, I always try to make things easy and reliable for the customer. Clear signage, fully stocked tables, and lots of steadily available, high-quality products. People are visual. Make it easy for them to see and understand what you're selling.

My motto at the farmers' market is, "If we don't bring it, we can't sell it." If a customer comes to market week after week, and I've neglected to bring a certain item he's been wanting, then this isn't easy on him. I won't get many more chances—maybe not even one more chance—before I lose that customer to the local grocery store, and never see him again. Make it easy for your customer to shop with you, and you'll make it easier on yourself to generate a profit.

Add Value

You're growing thousands of pounds of fresh produce every week, and spoilage and perishability will always be an issue. So will blemishes, bruises, and damaged products. Maybe you have a slow or rainy day, and now you're stuck with a truckload of tomatoes. Do you compost it? Donate it to a local food bank? Give it away to friends and family?

You certainly could. Or, since you already have access to markets and sales channels, you could buy a commercial dehydrator and produce "sun-dried" tomatoes—or, as we produce on our farm, liver jerky for pets (people seem to spend more on their pets than on their kids these days!). We end up selling a lot more liver snacks for Fido and Fifi than we ever would trying to market liver for people's dinner plates.

Value-added opportunities are everywhere. Do you have leftover apples, berries, cucumbers, cabbage, or herbs? Customers like jams and jellies, pickles and sauerkraut, apple sauce and cider, kimchi, pesto, and pasta sauce. Oh, and they really like pie, too!

Visualize the next logical processing step for whatever you're growing, and you might find an opportunity to add value to otherwise unusable products. This is especially useful regarding whatever crop you grow the most of, because economy of scale suggests this is where your greatest efficiencies already exist. If you can create a clever new product from this crop, you'll be unlocking additional value.

Adding a line of value-added products will require some additional capital, but the rewards could potentially come very quickly. You're already going to market anyway, so why not load up the truck a little more? Better yet, creating a value-added product line might be a great way to keep your farm crew staffed year-round, which is a problem for seasonal farmers, if ever there was one.

The number and value of Value-Added Producer Grants varied greatly between 2001 and 2015

Number of grants ■ ···· Obligated total funding, $ millions

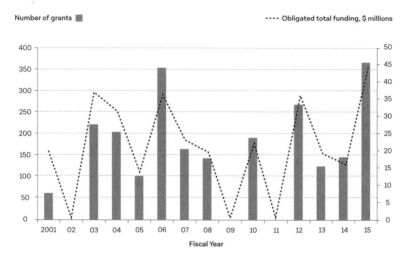

Source: USDA, Economic Research Service using data from USDA's Rural Business-Cooperative Service.

Give a Little Back

When something new is in season, offer free samples. Nothing works better to connect a customer to your farm than the smell and taste of delicious food, and doubly so if it's free. My sales typically increase by 25 percent if I just take the time to fire up a small propane grill and pass out samples of hot Italian sausage. For twenty dollars' worth of pork, propane, and time, I gross an extra four hundred bucks. Plus, I get to snack on any leftovers.

A few more simple give-a-little-back tips:

- Always round down on the scale a few cents. Or on a big order, shave off a dollar or two.
- For your best customers, occasionally offer a little gift—Cajuns call this a *lagniappe*—maybe an extra peach, or a bonus ear of corn, or the largest dozen of eggs.

- Schedule a day where customers are invited out to the farm for a tour. We have an annual potluck on our farm, where attendees either bring a side or a dessert, and we grill free hotdogs and hamburgers after the tour.
- Give away reusable shopping bags, calendars, refrigerator magnets, or a travel mug—anything branded with your farm's name that the customer can actually use while being reminded to shop with you.

What you grow will greatly shape how you market it. But ultimately, sales success will depend on how effectively you connect with your customers. Producing food for an appreciative clientele is truly one of the most satisfying aspects of farming. Find your markets, build your business relationships slowly, then follow your nurturing instincts. Tend to them with care. If you grow a superior product, folks will willingly pay for it and keep coming back for more. To have a thriving, successful farm, there's nothing more sustainable than happy customers.

CHAPTER REVIEW QUESTIONS

1. What are the unique marketing strengths and challenges presented by where you live, or where you'll farm? Create a list of each, then identify ways to accentuate each strength and turn each challenge into an advantage.

2. Think about your favorite local businesses. What is it about their locations, services, and products that you really like? How can you translate these qualities to your business?

3. Name your favorite food, and do your best to imagine the process involved in its on-farm production, harvest, storage, distribution, and transport. Finally, what sort of marketing must have occurred in order for it to have made its way to your table?

CHAPTER NINE

TAI CHI ECONOMICS

Ellen

There are two distinct sides to farm economics: First is your own proprietary metrics. Like in most businesses, you can organize farming finances into dozens of precise categories, applying proven accounting practices, budgetary strategies, and profit-and-loss statements so pristine they'd make a CPA giddy with delight. A farm business is all about costs, prices, incomes, yields, time cards, fertilizer rates, and even rpms. You have to become good with numbers, and you will. A clear understanding of the math, the finances, and the bottom line will always remain a top priority. Every successful, sustainable farmer shares a no-nonsense desire for clear economics.

But there's a second, more challenging aspect to managing a farm business: dealing with uncontrollable outside economic forces. Many of these are micro (such as the state of local economies, shifting sales demographics, or regional competition), while others are macro (such as government policies, national and global droughts, or the impact of international trade). Often these are factors that, despite your best efforts to adapt to them, could negatively affect your bottom line. Like it or not, what we're able to charge for tomatoes on our farm in Oregon is directly affected by the price of natural gas in Pennsylvania, or by the wages of underpaid farmworkers in Florida. Halfway around the world, an oversupply of grass-finished beef in New Zealand can immediately drive down the price of local hamburgers in Omaha, Nebraska.

It's certainly frustrating, but by identifying and navigating these external challenges as they arise, you can influence how they affect your business. Somehow, you must learn to leverage these headwinds to your advantage, relying on patience, intellectual dexterity, and lots of practice. This is what we call tai chi economics.

In tai chi, students are taught to meet incoming force with intentional softness and grace—not, as it might first appear, to escape or avoid conflict, but rather to engage the energy until it wears itself out. This not only conserves the strength of the practitioner but also allows them to observe and learn from their opponent. They calmly study the tactics and become more efficient in the future. To grow skilled in tai chi, the student must remain ever observant, engaged in the present, and subsequently meditative about the lesson.

Lao-tzu, the ancient Chinese philosopher, wrote, "The soft and the pliable will defeat the hard and the strong." Now if at first it seems strange to apply such a philosophy to farming economics, that's understandable. But cultivating a flexible, contemplative mindset not only helps you grow as a businessperson but also allows you to better enjoy the beauty of farming itself. Sometimes, it's useful to let the outside world slide past as you focus on the challenging work of growing food.

Which leads us to another useful proverb: "Those who say it cannot be done should not interrupt those who are doing it." As you grow your farming dreams, you will also grow your business acumen. Yes, you *can* make a profit while simultaneously growing nutritious food for the world. Yes, you *can* spend your days with your hands in the soil, yet remain an excellent accountant of your daily finances. In a world that's often binary, sustainable farming puts us on a path to something more holistic.

As we move forward in this chapter, we will provide mainstream economic definitions, alongside examples of how they apply to those daunting challenges that are unique to farming. So take a deep, calming breath, and let's explore the language and thinking behind some of these fundamental business concepts.

PROFIT

The most basic goal of a business is to supply a product or service in the marketplace, while generating a profit for the business owners. *Profit* is money that's left after all the costs of production, marketing, and overhead have been paid. Costs of production for this discussion should include every cost associated with the business, everything you write a check for, plus depreciation (a way to spread the initial cost of major equipment or infrastructure over its useful life). Here is a basic list of costs:

- Labor costs: all hired help, as well as your labor (the salary you need to be paid)
- Input costs: seed, fertilizer, feed, supplies
- Infrastructure costs: mortgage or lease payments, insurance, utilities, property taxes, depreciation
- Marketing costs: website, advertising, branding, trucking

Profits can be used in various ways, and the structure of your business will influence your choices. Like other business owners, you can distribute any or all of the profit as straight income (beyond paychecks). Having a great season and taking that extra income for your own pocket may allow you to help pay for your children's college tuition, take that dreamy tropical vacation, buy a hot tub to soothe your achy muscles, or even to get some expensive dental work done. And, of course, employees greatly appreciate surprise year-end bonuses!

You can also be wise and forward-thinking by putting some of your profits into retirement plans for yourselves and any qualified employees (more about this in chapter 10). You could choose to reinvest some of that profit in the farm business itself by making capital purchases or repairing major capital assets, such as a tractor, barn, or fence. Another good place to consider putting profits is into a rainy-day savings account. Most business advisors recommend having a cash buffer of at least enough money to cover a couple months' expenses, and up to a whole year's worth of payroll. Every one of these options has merits, and gets us excited. They entice us into wanting to make even more profits.

We shouldn't have to come straight out and say this, but *profit is good!* It's what allows a farm to grow, to pay its employees fairly, and to provide some security in case of an accident or disaster. Unfortunately, we find that many farmers seemingly have an allergy to the concept of profit. To them, "profit" is a dirty word—a goal that evil corporations go after relentlessly, wrecking lives and the environment along the way. But, it doesn't have to be that way. Here's your first tai chi opportunity—to make a profit while achieving the triple bottom line of a sustainable business: ecological stewardship, social justice, and economic viability. Echoing the three-legged stool construct (remember the three *E*'s—environment, energy, and economics), this diagram of three overlapping circles allows us to put our own spin on the word "profit" and make it ethically clean.

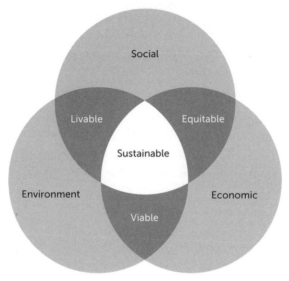

The Triple Bottom Line Approach

We suggest thinking about profit in a nonconventional way. Imagine that your farm has paid everyone a fair wage, your family's material needs are met, and your long-term assets are in good repair. What would you do if there were still money left over? Here are just a few ideas of what those profits might give you the freedom to do:

- Give away CSA shares to folks who can't otherwise afford your good, clean food.
- Attend a farmers' market in a food desert with your prices 25 to 50 percent lower than usual.
- Implement a more generous crop rotation whereby you actually grow on less land each year, resting more land than ever (which, by the way, will pay dividends later).

- Create a scholarship fund to help young farmers attend conferences or buy needed tools.
- Donate to your favorite charity.
- Work less hours by deciding to grow fewer crops next season and taking off one or two days a week all year long.

The choices are endless. And this, ultimately, is what profit allows—more choices. Choice is delightful. Choice is freedom. So as long as you are not taking unfair advantage of your customers or employees, profit is *good*. Let's reshape this antiquated idea of what profits are supposed to be about, and tai chi it toward positive inspirational goals.

ELLEN'S TURN TOWARD PROFIT

A pivotal moment in my career came during an honest conversation way back in 2003 with a farmer colleague from another state. I made the trek to visit her farm, and was staying the night. Breaking the taboo of talking about personal finances with anyone, I asked her how much money she actually put into her pocket each season. As in, what was her personal take-home pay that she could use to fund her life?

She told me her last few years of salary, and it blew my mind. Her number was twice what I was making! I had not thought it possible. Newly aware of what was truly attainable, I vowed then and there to increase my personal income. I came up with a new game plan for my farm, with profit-making now seated right next to our already established business principles of ecological stewardship, social consciousness, and working joyfully. Two years after that meeting, I had doubled my income. Making a profit is as much a mindset as it is a hard financial principle.

HOW TO SET PRICES

Of course, if you're ever going to make a profit, you need to know exactly how much to charge for what you grow. The equation seems easy enough:

Determine your production and marketing costs, add a percentage for your time and investment (we usually apply 25 percent here), and then set your price appropriately. If everything goes according to plan, those wonderful profits will soon start rolling in.

Sensible, right? The scary truth, however, is that most growers don't actually know what it costs them to produce any one product. It's unbelievable—an actual form of farming craziness. But this is the case for most of the non-commodity segment of agriculture—and we have a hunch that it applies to many commodity producers, as well. This must never be the case for you, however.

Now don't get me wrong. All growers know if they have any money in the bank at the end of the season. They know whether it was a "good" or "bad" season, both agriculturally and financially. But, almost no multicrop vegetable-, fruit-, or flower-grower can tell you *exactly* what it cost them to grow a carrot, a pound of apples, or a bunch of zinnias. And we're simply talking about production costs. We're not even mentioning a return on land investment, as discussed in chapter 5.

So why doesn't a farmer know what it costs to grow a single potato? This absurd situation is the result of a combination of factors. First off, multicrop non-commodity farmers grow a vast number of crops, usually between forty to sixty, sometimes even more. Many of those crops are grown in succession over the course of the season, for example, five plantings of summer squash (one every three weeks), or sunflowers seeded every week for twenty weeks. Further complicating things, often the same piece of land will produce more than one crop over the course of the season, and at rates of growth that might vary from one year to the next. An early lettuce crop might be followed by green beans, which is subsequently followed by fall-planted garlic.

Those are the agricultural reasons. The other huge complicating factor—and the biggest operating cost of all—is labor. Believe it or not, it's difficult to keep track of how long every person works on each and every crop, each and every day.

Here's a typical scenario on any market farm. On a Friday morning, a group of four people will set off with a list of ten crops to harvest and then wash, pack, and put into the cooler. Starting in the chard patch, they spend thirty minutes there, harvesting. Then, one person heads off to pick escarole for ten minutes, while the other three cut heads of lettuce. Sometime later, the escarole-picker rejoins the lettuce group, helping to finish. Then, all four are off to the herb patch for forty-five minutes, where they harvest twelve different herbs. With a truckful of produce, one person heads back to the barn to start washing, while the other three move on to pick the summer squash patch.

You can see, in this example, how numerous crops are worked on for varying amounts of time by varying numbers of people. And that's just a single three-hour scenario. Tracking labor by crop is hard! Thus, almost no growers even attempt it. A good grower can certainly report how much he paid for labor in total, and then what percentage of total farm expenses the labor comprises. If they're really sharp, they can say how much each dollar of labor produces in sellable products. But none of that helps them know *precisely* what it cost to grow a single tomato.

Simply put, pricing based on real costs on a diversified farm is a complex process. Short of doing the hard work of tracking labor and production cost by crop, the best you can do is to keep increasing efficiency (more output per unit of input), and to grow better-yielding crops. These are certainly worthy goals, and we highly recommend them both. But what remains elusive using those tactics alone is whether there are crops that are consistent money-losers, ones that should be dropped from the list. There might also be crops for which you could profitably lower the price, in order to gain more retail market share or to access a lower-priced wholesale market channel.

Yes, it *is* possible to use individual crop budgets (also called enterprise budgets) for every crop you grow. A crop budget is a template that attempts to distill all the costs and income for a single crop, so that any grower can plug in their farm's numbers to see in advance if growing that particular crop might be profitable. Land grant institutions around the country have developed and made available many crop budgets. You can use these budgets to examine the cost structure of any given crop, but you still have to have good data in order to end up with a meaningful result—the *real* cost of production of that crop. And that leads us back to labor hours.

If you want to know exactly what it costs to grow each and every crop, you would have to create as many budgets as you have crops—this could be an unwieldy forty to sixty budgets! Another problem is that individual crop budgets are less rigorous at accounting for your total farm overhead costs. On top of this, they don't shed light on whether the cost of selling through one market channel is different from another—for example, wholesale versus retail, or CSA versus farmers' market.

So how *do* growers set their prices? The most common strategy is to look around the marketplace and see what everybody else is charging. A confident farmer charges a little more, or if they aren't confident about the quality of what they are offering, they charge less. For commodity or wholesale producers (discussed in chapter 7), the producer often has to simply take what the market or auction happens to be offering that day. In both instances, the price is fixed to a narrow range of numbers, largely affected by outside factors—factors that have nothing to do with what it cost you to actually grow the crop!

The flaws of these systems and strategies are clear. The price has essentially no relationship to what it *truly* costs to produce and market the item. In essence, these methods depend on the market—other growers—to look out for your bottom line. The problem, of course, is that none of those people give a hoot about your bottom line. Why would they? None of those competitors in the market know what their costs are either, so now perhaps everybody is losing money on carrots.

This is one crazy marketplace, and this sort of blind pricing happens all the time in farming. It's difficult to be a true maverick, setting your prices to reflect your full cost of production. Never forget that consumers have been trained to expect a certain ceiling on what food should cost. When you try to set your prices independently, which often results in a price hike, there's usually customer pushback.

This is all context to show why you need to employ tai chi tactics. Instead of following your fellow growers and accepting the prevailing price, you must be smart enough to let this faulty strategy flow around you with little concern. In other words, you must determine what price *you* need to stay in business.

We know it's easy to say, "Set your prices on the high end" or "Grow lots of one crop efficiently, and price it competitively," and simply leave it at that. But, unfortunately, nothing is ever that straightforward. Using tai chi economics means that you acknowledge that pricing pressures can vary widely, depending on your local community, region of the country, and access to markets. Generally speaking, higher prices are more reliably achieved with access to metropolitan markets, and much more difficult to sustain in a small, rural community.

So how will you determine your pricing? You'll need to adapt accordingly—but always striving toward the highest prices that the market will bear. With that in mind, here are some price-setting strategies from poor to best:

- Poor: Look around the market, especially at grocery stores, and think about what you offer that is better or different than your competition. Price accordingly.
- Good: Run some numbers using enterprise budgets on your major crops, or look at publicly available crop budgets. Be careful to understand what the budgets are assessing and what definitions they use for profit, overhead, and owner's labor. Try to figure out what your cost to produce is, and price accordingly.
- Better: Track labor and every expense associated with your top five crops (designated by most acreage or highest sales), and see what the total cost to grow, harvest, and market them is. Price from there. Don't forget to pay yourself in the equation!
- Best: Track all labor and expenses on the entire enterprise (all crops), including all overhead and your own pay, run it all through Veggie Compass (see below) or a similar program to see *exactly* what your break-even price is for each and every crop in all of your market channels.

The Veggie Compass Program

Veggie Compass is a free Excel spreadsheet program from the University of Wisconsin–Madison, created by Jim Munsch and John Hendrickson. Veggie Compass, at the time of this writing, is the only tool available to help vegetable growers compute their own true costs of growing up to one hundred crops. It is a rigorous program that takes into account *all* the costs of the farm, including overhead, depreciation, and marketing. Currently, a number of new farm-management software programs are being tested and will hopefully be available to farmers in the near future. Jim and John are poised to release new versions of their program—Protein Compass for livestock producers, and Nut and Berry Compass for those growers with perennial crops.

Veggie Compass— Whole Farm Profit Management	FARM PROFIT &		
	CSA	Farmers' Market	Wholesale
Sales—by Market Channel	$39,000	$210,490	$91,100
Market Channel sales as % of Total sales	9.83%	53.08%	22.97%
All Production Expenses	$20,703	$109,687	$59,035
Production Expense as % of Mkt Chan sales	53.08%	52.11%	64.80%
Gross Profit	$18,297	$100,803	$32,065
Gross Profit as a % of Mkt Channel sales	46.92%	47.89%	35.20%
Market Channel Expenses	$6,005	$24,569	$3,456
General Mgmt & Admin Expenses Allocated to Market Channel	$11,278	$60,867	$26,343
Total Market Channel plus General Mgmt & Admin Expenses	$17,283	$85,436	$29,799
Total MC & GM Exp as % of Mkt Chan sales	44%	41%	33%
NET PROFIT	$1,015	$15,367	$2,265
Net Profit as % of Mkt Channel Sales	3%	7%	2%
Net Profit as % of Total Net Profit	9%	142%	21%
Non-Operating Income			
USDA Program Payments			
Patronage Dividends			
Interest Income			
Other Income			
Taxable Income			

One of the first users of the Veggie Compass program was rewarded with a new powerful take-home message—he was making a healthy profit on a number of storage root crops. Apparently, he was very efficient at growing these crops, with great yields and low costs of production. With that new knowledge, he scaled up production of those crops, and could then profitably sell more of them via a wholesale channel. These root crops became an essential profit center of his business. Without that detailed analysis, he would never have known this was possible.

LOSS BY MARKET CHANNEL

Stand	Total from On-Farm Production	Buy-Resell	Grand Total
$26,610	$367,200	$29,370	$396,570
6.71%	92.59%	7.41%	
$14,591	$204,017	$16,280	$220,297
54.83%	55.56%	55.43%	55.55%
$12,019	$163,183	$13,090	$176,273
45.17%	44.44%	44.57%	44.45%
$8,360	$42,390	$8,349	$50,739
$7,695	$106,182	$8,493	$114,675
$16,055	$148,572	$16,842	$165,414
60%	40%	57.3%	42%
-$4,036	$14,611	-$3,752	$10,859
-15%	4%	-12.8%	3%
-37%	135%	-34.6%	100%
			$0
			$0
			$0
			$12,000
			$22,859

Downward Price Pressure (How High Can You Go?)

Outside of the traditional commodity markets, you can ask for any price you want. But, that doesn't mean that a customer will agree to pay it! You have to employ some level of reason and try to delineate what is obviously too much to ask. Again, it's time for some tai chi economics.

Take the price of a dozen eggs, for example. In the mid-Atlantic, we can find a dozen white eggs, packed in a Styrofoam carton from an essentially anonymous, industrial-model farm, at a QuickMart for $1.99. At the extreme other end of the spectrum, we can find a dozen brown eggs, packed in a paper box with a gorgeous color label, that came from free-range chickens fed organic grains, grown on a local raptor-friendly farm, for $5.99. That's a price spread of $4 dollars, more than a 200 percent difference. That's a wide price range for twelve eggs! But how high might a customer be willing to go? Will some people pay $10 per dozen? Maybe. But will someone pay $15 per dozen? Absolutely not. On our end, we know that we must make more than our costs—and preferably, as much as the market will reasonably bear. With these numbers in mind, we at least have some general pricing parameters to start with.

Let's say, for this example, that your break-even price (the price that covers your costs but doesn't make a profit) for eggs is $7. You have a few options. You can try to sell them at $8 for a profit, and see what happens; you won't know for sure until you try. You can sell them for $7, and break even. Or you can sell them for $6, like most of your farmers' market competitors, but lose money. (If you ever find yourself in this last situation, it's much easier and more efficient to simply hand out dollar bills at market than to go to the trouble of keeping laying hens and selling the eggs for a loss—trust us!) Another option is to figure out how to lower your cost of production, so that you can potentially make a profit at $7. Or you can refrain from keeping laying hens altogether.

This is a pretty powerful position to find yourself in—having this many choices is wonderful. So while the prevailing notion is that customers "will not pay this or that price," we know different: For most foods, there is still plenty of wiggle room at the top end of the price range. It's plain to see that this price-setting job is a never-ending feedback loop of determining what exactly you are selling, knowing your costs, and seeing what price you can get in real time—a price that's somewhat influenced by external market factors beyond your control.

From a tai chi point of view, this means several things:

- People will gladly pay for quality, superior taste, consistency, and a meaningful story. If you grow a superior product, then you must fearlessly set your prices accordingly. This usually means pushing the upper limits of what a customer is willing to pay.
- You must cultivate a marketplace that will support these prices (see chapter 8).

- If you are growing a unique and superior product and marketing it correctly, technically, *there is no competition*. This means that your product becomes a one-of-a-kind, and you can charge more for it.
- Finally, never apologize for any of this! No other business makes excuses for supplying a high-quality product and asking a fair price. Neither should farmers. Let this antiquated twentieth-century notion of low food prices flow right past you, while you concentrate on growing a great product and generating a profit. Your goal should be to remove all emotion from pricing. It's a living, flowing system, one that is well represented by the tai chi symbol, where you are the dot.

POWER AT THE BARGAINING TABLE

Another grower I know who worked with Veggie Compass took his newly discovered break-even prices to his annual meeting with Whole Foods. He showed the buyers his spreadsheets and successfully argued for a higher price for his produce. In the past, their conversation had involved him asking for a better price and not getting it. But this time, he turned that around by telling the buyers the price he needed to remain profitable. With the data on his side, he got the price he wanted. Honesty, transparency, and insistence on a well-earned profit—this is tai chi economics in action!

NITTY-GRITTY FINANCIALS: ACCOUNTING AND BOOKKEEPING

No business runs without keeping books and performing basic accounting practices, and farming is no exception. Bookkeeping is the recording of all financial transactions: sales, expenses, and banking. Accounting is taking that financial data and organizing, analyzing, summarizing, and reporting it. Think of the "books" as the numbers themselves, and accounting as the story of what those numbers represent, giving them context and meaning.

Many farmers do their own bookkeeping. I did it for my farm for twenty-five years, and I enjoyed the work. My experience is that the very act of inputting every expense, every credit-card purchase, and every deposit kept me strongly present to the financial side of my business. This is a tai chi state of mind.

There are plenty of farmers, however, who outsource bookkeeping to a professional. The upside of paying someone to do the books is that the job is usually done well and on a timely basis. The downside (apart from having to pay for the service) is that those numbers are less familiar to you, less visceral, and you are more likely to miss opportunities—or warning signs—that might otherwise be clear.

On a regular basis throughout my farming career, I would see the bank account balance looking meager and wonder, *Where did all the money go?* Because I did the books myself, I could sit down with my QuickBooks program and get an immediate answer—"Oh yeah, this was the season for all new greenhouse plastic," or "We had to replace the well pump in February." It's amazing how quickly I could forget past purchases! Having the data at my fingertips gave me timely information and eased my mind.

If we are the masters of our own finances, the answers are only ever a click away, with no professional assistance required. So despite the common wisdom that a small business needs a hired bookkeeper, the tai chi stance is to embrace the challenge and be the champions of our own numbers. We recommend that you do your own books, always keeping them separate from your personal and family finances. The trick is to get some training, and to schedule a regular time each week to work on the books.

Using a program such as QuickBooks allows you to generate meaningful reports about income and expenses, including the useful feature of comparing one season to another. But a professional accountant can take your books and help you think more strategically about your finances.

For example, perhaps you want to build a new shed to house the wash-and-pack stations for your produce. You need to think through how to pay for this capital expense—whether to borrow money or use cash, or some combination of both. An accountant can help coach you through that decision. (Believe it or not, using cash is not always the wisest course of action.) Additionally, an accountant can prepare your taxes, which is a huge relief.

You need an accountant who is conversant about agricultural businesses. As with all your relationships with service providers, it's important for you to be honest and comfortable. You need to be firm but friendly, and insist that the language and terminology used in your bookkeeping is meaningful to you. For example, an accountant would be satisfied with your having one giant category of expenses called "supplies." But for you to manage your business well, it would be more helpful to make subcategories to track specific sets of supplies—such as "greenhouse," "office," "marketing," or "veterinary supplies." So, just like when you visit the doctor, you shouldn't be shy about telling your accountant the whole story. Remember that the accountant is working for you, not the other way around! As long as you have a good working relationship with your accountant, the money you spend on their services will be well worth it.

If you find yourself needing to intersect with commercial lenders or mortgage companies, then you will have to take the time to learn about financial statements, cash flow analyses, and financial ratios. Your accountant can take your well-kept books and help pull your head out of the numerical sand to do these higher-perspective business analyses.

On my farm, in the years before we began selling via the CSA model, we always had a dearth of cash during the lean sales months of March through

SHOULD YOU REPORT ALL YOUR EARNINGS?

A note on *not* reporting earnings on your taxes. When running a cash-based business, it is tempting to hide some income from the IRS. The upside to this is obvious—no income tax to pay on the hidden cash. The downsides are harder to discern. The less you earn as personal income, the less deduction you pay for Social Security and Medicare. That means that you will qualify for significantly less Social Security income when you retire. Fudged income numbers also paint a weaker picture of your farm's financial prowess. Should you ever need credit or to borrow money for expanding your business in the future, your profit-and-loss statement will be less impressive and may preclude you from obtaining a useful business loan. We recommend reporting *all* your farm income.

May. Our remedy was to get an operating loan from Farm Credit. Basically, this was a line of credit, almost like a credit card. But instead of making monthly payments, we were able to set the payback date as far out as six months, when we were flush with cash. This, once again, is using tai chi to flow around the conventional system. Instead of being forced to conform to mainstream financing, we found a lender who worked with *our* time frame. Now we could simply focus on doing what we did best—growing our crops and our profits within the growing season.

This loan application required both a profit-and-loss statement and a balance sheet, and working with our accountant made us much more comfortable with the process. We recently went through the same scenario to obtain a mortgage for purchasing a house to rent out to employees. The process is similar to buying a home for ourselves, only this time, the lender needed to see if the business would be able to pay them back. Those statements and sheets told the story of how much we could afford to borrow.

DEPRECIATION CAN BE A VALUABLE TOOL

Depreciation is a way of dealing with the original cost of major equipment and infrastructure that will last for five or more years. Such assets provide financial benefit over their useful life, but also depreciate in value through wear and tear. At some point, the item will no longer be usable and will have to be replaced. Current tax law generally allows for deducting the full cost of these assets in the year they are procured. While this is good for taxes, this does not match the expense of the asset with the benefit derived from it over its life. Therefore, using tax-driven depreciation will introduce a potentially significant error in determining the cost of production.

A more accurate assessment of the yearly cost is to use some form of traditional depreciation. A simple method is called "straight-line depreciation," whereby the initial cost of the asset is spread evenly over its useful life. For example, if you were to purchase a tractor for $25,000 this year, the IRS would consider this an asset having a five-year useful lifespan. Using the straight-line depreciation method, you would depreciate it at $5,000 per year for five years. This would allow your business to bear the cost of that asset over time, instead of taking the whole expense off of this year's taxes. We suggest further study *and* employing the services of a certified public accountant to complete your business taxes.

YOU CAN'T MANAGE WHAT YOU DON'T KNOW

A wonderful agriculture business guru of mine introduced me to the saying, "You can't manage what you don't know." At first, I thought, *Well, of course you can't*, and briefly left it at that. But the more I chewed on it, the more complex and nuanced this idea became. Applied to farming, this saying directs us to know as much as we can about what is actually going on, out there on our farms. Channeling your tai chi demeanor means keeping your senses awake. You not only need to know what is so; you also need to record it. Farming is as much a management game as it is a technical workout. If you don't know what is what, then you can't learn from your mistakes and make positive adjustments.

And . . . bingo! I've tricked you into thinking about record keeping! It's merely the act of writing down what happens on the farm. In order to make improvements in your production and marketing, you need to know what you've done before, and how well it worked.

Let's say that you want to try a newfangled lettuce variety for a salad mix, one that the seed company claims will increase yield and cut labor costs. If you don't track the pounds of lettuce harvested and the time it took—as well as what your prior harvests have been—you won't know if spending triple the money on seed is worth it.

Simple, right? But not unless you actually keep records, instead of just agreeing that it's a good idea. Luckily, you don't have to do this ongoing exercise alone—everyone who works on the farm can take part. It should become a normal part of your farm's culture. "We keep track so we can keep improving" is a great mantra.

There are external reasons to keep records as well—such as for certifications of any kind (organic, biodynamic, certified humane), and especially for food-safety compliance. Here are some key areas where keeping records is fundamental:

Seeds: variety, source, amount, cost

Livestock: birth dates, wean weights and dates, tag numbers, sire and dam

Equipment: model number, serial number, engine hours, cost, maintenance schedule, repairs

Crop rotation: what got planted (where and when), acreage, plant population (spacing)

Field notes: how different crops performed, insect and disease issues, weed report

Yield: harvest amounts and dates, and source (which field or patch)

Sales: wholesale invoices, farmers' market reports (what went, what sold, prices, weather, comments), CSA sales, and contents of each week's share, farmstand sales report

Fertility: what was applied where, amount, timing, source

Labor beyond payroll: activity type (growing, harvesting, selling) on what crop. (The short-term goal is to create your own farm metrics. For example, it takes five person hours to plant one hundred bed feet of garlic, or it takes thirty minutes to wash fifty bunches of beets. The long-term goal is to arrive at the ever-elusive cost to produce crop X.)

Records can be kept using myriad methods. Good old pencil and paper still works well for many growers: log books, wall calendars, day planners, and spiral-bound notebooks. Erasable whiteboards are great for tracking daily activities and yields, which can then be digitized by taking a photograph, or by regularly inputting the data into a spreadsheet. I grew fond of keeping track of my major farm activities using a digital cloud-based calendar: tillage, fertilization, seeding, transplanting, first and last harvest. I could search the calendar for every time the word *beet* was used, or the name of a field, such as "Shed Patch," to see what happened in past seasons, and on what date. These calendars can be easily shared with workmates and partners.

Taking photographs of all kinds of farm activities is an easy way to record information. I know a grower who uses her Instagram photos to look back at what kinds of products were harvested and what activities happened on which days the previous season—a clever way to organize a pictorial memory bank.

Choose from the various record-keeping methods, and pair them to your particular strengths, habits, and routines. *But keep good records!* Record keeping will never be the reason anyone goes into farming, but those records will help make you more prosperous and efficient.

SENTIMENTAL RECORDS

When I handed over the reins to my farm this past spring, one of the most sacred and important records to pass on was a big green-tinted ledger. It held a concise record of what had been planted in every field for the last twenty-five years. I had referenced that book every winter, to figure out where to plant the next season's crops. Seeing the farm's agricultural history depicted on these big sheets of paper gave me such a feeling of accomplishment and pleasure. What an amazing distillation of my years of work.

Tai chi economics is all about leveraging the methods and madness of the mainstream economy to your advantage, while leaving the clichés behind.

Articles in our national newspapers tell us that farms are struggling and going out of business at an alarming rate, that suicide among American farmworkers is high, and that corporate agriculture is gobbling up farmland. The messages can be downright discouraging. In order for you to succeed, you first have to acknowledge these warnings, and then use all your skills, knowledge, and creative energy to work around the obstacles. Sure, you'll employ the conventional tools of bookkeeping and financial management, but you'll also develop your tai chi grace, allowing non-useful energy to flow right past you on your way to a profitable, financially organized, and economically sustainable farm.

CHAPTER REVIEW QUESTIONS

1. What is profit, and what might it do for you, your family, and your business?

2. How will you set prices?

3. Will you keep your own books? If so, how will you learn to do it? If not, who will do them for you?

4. What record-keeping systems will you create? Who will maintain those systems? What is the purpose of keeping those records?

. CHAPTER TEN .

MAKE A REAL PROFIT,
SECURE YOUR FUTURE

Forrest

The amount of money that can be spent on a farm is potentially endless. And we're not talking about trivial finances here. We mean significant, justifiable everyday expenses—the bare minimum that's often required to make a farm operate each day. Fences must be maintained, and trucks serviced. The fans in the greenhouse must run, as well as the compressors in the walk-in freezer. Drip tape has to be replaced; the barn will need a new gutter; the muddy road requires fresh gravel; the trusty old shovel, finally splintered beyond repair, will have to be purchased anew. You woke this morning to discover that your majestic, centuries-old oak tree finally succumbed to gravity . . . and fell onto your neighbor's property. Better hope your $300 chainsaw fires up, or you'll be paying $2,500 to a tree service company.

Of course, this is all just in a day's work, the predictable grind of maintenance and repairs. But we haven't even mentioned important expenditures involved in improvements and advances—installing new watering systems for your livestock, expanding your loading dock, revamping the washing station for vegetable processing, doubling your milk tank capacity, overhauling your irrigation system—investments that eventually improve your cash flow and bottom line. Oh, and did we mention the money required for taxes, workers' compensation, health insurance, marketing, and payroll? How about that all-important (and elusive) paycheck that you write to *yourself*, as well as the one to fund your retirement savings account?

Let's be perfectly clear. Even though daily expenses never end on a farm, there will always be reasons to spend even more—and for perfectly logical, revenue-generating reasons. It makes sense that you would want to make an improvement or create an efficiency if you can. This is the way all businesses get ahead, not just farms. But especially when it comes to agriculture, there are so many excellent and logical reasons to justify spending your savings that the actual accomplishment, often, is *not* spending money! It seems silly to have to say it, but cold cash will burn a hole straight through your Carhartts. I'd be embarrassed to recount the number of times I've spent an extra two or three hundred bucks at Tractor Supply Company, just because I had it.

Profitable farming is already a high-wire act. So at the end of the day, it's your job—your responsibility—to take the important, disciplined steps of using your profits to build a financial safety net to catch you when you stumble. And have no doubt: You *will* stumble. This means that you must become a devoted budgeter, conscientious saver, and educated investor. *A safety net always has to be there.*

In chapter 9, Ellen discussed how to set your prices, and what it really means to make a profit. Now you need to understand how much you should budget, and how you can best secure and grow your hard-earned money. As is so often the case, the answers can at first seem counterintuitive. Let's start with examining this diagram:

THE MONEY TRIANGLE

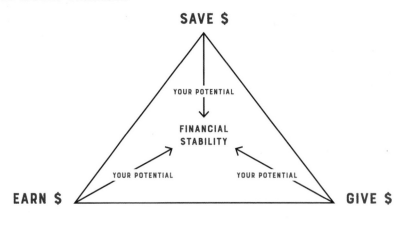

Theologian John Wesley once wrote, "Earn all you can, save all you can, give all you can." I've always been extremely fond of this expression, because it accomplishes so much, so succinctly. Applied to farming, it guides us, simultaneously, toward economic and philosophical prosperity—a very noble goal! Yet at the same time, it actually doesn't seem to explain anything at all, at least not precisely. Studying the diagram more closely, it almost seems self-nullifying: How is it possible to accomplish all three of these goals at once?

It's worth your time to regard this diagram as a mathematical equation. The three arrows point toward the center, where balance is eventually achieved. The trick here is understanding that each arrow can be extremely variable, using metrics that apply to the individual. With the phrase "all you can," Wesley cleverly superimposes personal capabilities and individual resources over something as deeply subjective as money management. In this model, there's no specific dollar amount. It's all about potential and possibility—specifically as it relates to you.

Of course, the first thing to catch your eye is probably "Earn all you can." Nothing wrong with that, especially in the difficult business of agriculture! But this turns out to be a bit of a red herring. What's the point of earning money, ultimately, if you can't save some of it, or give some back? On a farm, you might make money all day long, yet spend it just as quickly. But if you can somehow manage to save some of that money, it gradually helps you to make even more. Savings, it turns out, are the catalyst in the flowchart for financial success:

Savings Flowchart

Savings builds security → Security builds confidence → Confidence allows us to take risks → Risk-taking is where money is made

But how is it possible to make and save money if you're giving it away at the same time? This, again, is where the critical phrase "all you can" comes into play. We've repeatedly emphasized the importance of giving back to your soil, your employees and families, and even to yourself, but whenever you give back to your community or to a charitable cause, the good deed is almost always repaid in one form or another. Don't ask me how or why, but from our experience, it just is. This is one of life's great mysteries, and it's a sure sign of a well-functioning, sustainable system. Making, saving, and giving are all inextricably linked, three angles of a triangle that support each other.

The Magic of 10 Percent

This all sounds pleasantly philosophical, but it begs a more tangible question: Isn't there a specific financial rule when it comes to earning, saving, and giving? The short answer is *maybe* (and hopefully by now you're appropriately skeptical of financial generalizations). My personal framework is what I call the 10 percent plan. Let's say, for example, you gross $100,000 in a year. Here is how the plan would work:

- Plan to make at least 10 percent net profit ($10,000).
- Plan to save at least 10 percent of these profits ($1,000), placing this money into investments that subsequently make approximately 10 percent ($100).
- Plan to give away at least 10 percent of what your investments make for you ($10).

Why 10 percent? Because of human nature. This number is the Goldilocks of economic percentages: not too hot, not too cold . . . but just right. The world is filled with people trying to get rich quickly, forever chasing hot new moneymaking ideas, known as bubble economies. What always happens with bubbles? They pop! Conversely, some people—many, many people—are so risk-adverse or set in their ways that they never invest in anything. Or even worse, they spend their entire lives mired in debt. But 10 percent is a perfectly attainable number for most to achieve—a dime out of every dollar made, a penny from every dime subsequently placed into investments, and then 10 percent again given away from what those investments create.

It is also an ideal percentage to use for the maximum amount of debt you should accrue during the course of a year: Never borrow more than 10 percent of your annual gross income, and certainly never borrow at a rate of more than 10 percent interest—in fact, no more than 5 percent, if you can. This is also why you must pay your credit cards off in full every month. The average credit card charges more than 15 percent interest! Keep thinking *10 percent* at every step of the way, and the math will gradually become very favorable for you.

To be clear, these percentages are minimum goals. What you actually make, save, and give will all depend on the "all you can" factor. There are many years when I make far more than 10 percent net profit (net profit is your final earnings after all production and operating costs have been accounted for). In fact, over time, 25 percent has become my normal goal, and it makes my farm run like gangbusters, when I'm able to achieve it. Very occasionally, I'll even make a 30 percent return, which not only helps offset my land investment costs but also sometimes even allows me to consider buying *more* land—and under this specific economic scenario, I have done so.

Have no illusion, however, that a return of 10 percent is easy to achieve. Even though I plan to make 25 percent each year, there are almost always unexpected costs that pop up, forcing me to become more efficient or create an extra sales channel in order to achieve this goal. After all, most Fortune 500 companies average 8 to 10 percent net profit margins over the course of their business trajectories. Clever as I try to be, I seriously doubt that I'll outsmart five hundred successful mega-businesses over the full arc of my career!

That said, as long as you set an average expectation of a 10 percent profit, you've at least established a rock-solid benchmark. It provides a reasonable, achievable goal over the long haul, and will gradually provide consistent, robust, and sustainable growth. The next question, of course, is how do you ever manage to save money on a farm?

THE WAVE AND THE CUP

We've mentioned the all-important concept of cash flow several times. If everything goes according to plan, your production and sales should generate a reliable influx of weekly cash, and perhaps even lots of cash. A steady cash flow is incredibly important for paying your bills on time, writing checks for payroll, and funding budgets.

On my farm, each weekend at market, hundreds of pounds of ground beef, sausages, and steaks are exchanged for many thousands of dollars. When Monday morning rolls around, it certainly seems like a lot of money! But by the time my bills are paid, it's uncanny how quickly all that money has disappeared. Uncanny, and also discouraging, like a wave of cash crashing over me, then flowing back out to sea. I often find myself wondering, *What in the heck happened to all that money I just made?*

I call this the wave and the cup phenomenon, where I stand on the beach each weekend, holding a little tin cup, while a tsunami of dollar bills crashes over me. What lands in the cup is what I'm able to keep, while everything else goes back out to sea. Now if this sounds a little pathetic, that's okay. Perspective is everything, and there were many years that I didn't even have a cup! As long as I can capture some of it, I can gradually make progress. It's vital to understand that this is a long-term strategy.

Remember how this chapter started: There's no end to the amount of money that can be justifiably spent on a farm. If you think that you'll be the one who breaks this rule, please know that that's never been the case for me, or any farmer I've ever met. Instead, you must commit to the financially disciplined, and admittedly very difficult, position of budgeting your cash flow and sequestering that wave of dollar bills while it lingers—however briefly—in your cup. You've heard of get rich quick? This is get rich slo-o-o-o-ow. But it's important to know that you *can* grow your wealth. It all starts with organizing your priorities and assembling a budget.

CREATE A BUDGET THAT WORKS EVERY TIME

I've seen no shortage of strategies over the years about how to budget for a business. While I'm sure that they all have their merits and rationales, trying to follow someone else's budgetary recommendations has never worked for me, especially when it comes to farming. Each enterprise is so different, it seems illogical to try to apply a one-size-fits-all solution.

Instead, I suggest the following loose framework. Sit down with a cup of tea, a pen and paper (or a laptop and spreadsheet), and, including every minuscule detail that you can possibly think of, make three distinct lists:

1. Recurring expenses: biweekly (payroll and weekly operating costs), monthly (utility bills), and annual (taxes) expenses, plus 10 percent of your weekly revenue to be invested

2. Major operational necessities: paying for a tractor or a barn
3. Major operational replacement allocation: replacing a tractor or a barn

Next, you need to perform some simple math. Assign an estimated dollar value next to each entry, then divide that by the respective time frame. Again, be as specific and detailed as possible with the recurring expenses, from gasoline to chicken feed, and all the way down to twist-ties for the lettuce bags. For example, if you know you'll need $2,000 biweekly for payroll, that's either an intimidating $52,000 per year, or a relatively benign $1,000 weekly budgetary allocation. Similarly, if your electricity bill averages $1,500 a month, that's either a worrisome $18,000 annually, or a far more manageable $346 each week. With cash flow often varying widely from one month to the next, having these numbers broken down into weekly allocations can be just as important for maintaining your personal morale as for meeting your annual budget.

Add all these numbers up, and it gives you a very clear picture of what sort of weekly revenue you must generate. Do you need to gross $2,000 each week, or is it closer to $10,000? These lists inform your decisions regarding production, marketing, and staffing. Far better to know exactly how much cash the farm must bring in than to find yourself holding an "empty cup" at the end of the week.

What about major operational necessities, like that $65,000 tractor you've been eyeing at the dealership along the highway? You know, the shiny one with all the horsepower and cool attachments—the one you've convinced yourself you'll sadly never be able to afford. It's true, you'll never be able to afford it if you never budget for it. To be sure, we always recommend first renting or outsourcing any heavy equipment for your first five years of operation. But there's no reason that you can't also be budgeting for your own tractor along the way.

Whereas $65,000 might seem impossible in one lump sum, spread it over five years, and it becomes a much more manageable $250 per week. Better yet, if you develop pristine credit—as you very well might, adhering to this system—be patient and wait on a year-end deal. If you can leverage that $250 per week over an additional five years at zero percent interest, you'll effectively spend just $125 per week instead. This is exactly how I bought my first tractor, paid off monthly with money I had budgeted well in advance. Not painless, exactly, but not *painful* either. With careful use, I expect that tractor to last at least twenty years.

You might be wondering about the major operational replacement allocation list. This is a really important one. As you have probably realized by now, when it comes to farming, nothing is quite as valuable as peace of mind. With all the variables of weather and seasonality, it's imperative to know that, when something goes wrong, you'll have the resources to quickly fix the problem. Of course, the most important resource is your pluck and ingenuity. But growing this budgetary fund—and the peace of mind it affords—is really helpful as well.

Nearly a decade ago, I remember the sinking feeling of my market truck dropping out of gear along the highway, the engine still working, but the transmission nonresponsive. After four hours of waiting for a wrecker big enough to haul it, I rode shotgun to the repair shop, where the mechanic delivered the diagnosis: The transmission was completely destroyed, and a replacement was no longer made for my vehicle. Short of scouring Eastern Seaboard junkyards for a 1993, four-cylinder manual Mitsubishi Fuso transmission, my truck was effectively defunct.

Fortunately, I had gradually squirreled away money precisely for this potentiality. One week and a thousand Craigslist search results later, I found a used catering truck with less than ten thousand miles on it, for $15,000. I offered them $11,000 in cash, and they accepted on the spot (in hindsight, I wish I had offered $10,000!). As of this writing, ten years later, that truck still carries my products to market each weekend.

THE MOST STRAIGHTFORWARD BUDGETING STRATEGY YOU'LL EVER SEE

So how do you go about setting aside this weekly money? There are lots of ways. You could establish an electronic banking transfer, hire an accountant, or subscribe to an app that automatically manages your receipts in real time. This is the way lots of businesses operate, and that's fine.

But many farmers—myself included—still operate with cash sales. If you find yourself generating a sizable pile of twenty-dollar bills each week, then I suggest simply copying what I do: Buy a used, fireproof, four-hundred-pound safe for fifty dollars, a box of oversize envelopes, and methodically allocate cash into budgetary categories 2 and 3 (major operational necessities and major operational replacement allocation). Call me old-fashioned, but hey, this system is remarkably effective.

Of course, with more and more electronic (noncash) sales transactions, you might have to set up individual accounts and carefully move your money each week. Like most things on a farm, all it takes is actually doing it. But regular budgeting takes discipline, especially at first. Stick with it until it becomes routine, and your successful routine becomes a tradition. Seeing genuine profits at the end of the year is really motivating. With practice and intentionality, I've found that budgeting becomes a self-perpetuating mechanism.

In this system, it's only after you've squirreled away money for major operational necessities and their replacements that you dedicate whatever cash remains to recurring expenses. Sticking to this plan, the finances gradually become self-regulating.

Best of all, if there's money beyond what's been budgeted and spent that week for recurring expenses, then you actually might have cash left over to buy whatever you want. What a great feeling! This is known as discretionary spending—buying what you want to buy, as opposed to buying what you must buy. This is the most satisfying point of all, and it's the moment when you know that things are really clicking.

All of that said, if discretionary money is not there, then it's no huge deal. At least your weekly and long-term expenses are accounted for, and you will sleep well with peace of mind. Once again, *there's no end of money that can be justifiably spent on a farm*. This is why it's imperative to stick to a budget, trust in it, and gradually build upon it. After decades of farming, this is the system that works best for me—and I've tried a lot of systems over the years.

A final note about debt. Recurring expenses is also the place where you will pay back any prefarming debts you might have (e.g., student loans) or accelerate their repayment through discretionary spending. A good rule of thumb is to pay back all prior loans as aggressively as you can. Until they are retired, old debts will hamper your ability to invest in your farm, and loan interest will continue to accumulate in the meantime.

INVESTING OFF THE FARM

There's an important fourth budgetary item (technically, a subset of recurring expenses) that becomes available once the first three are established: investing off the farm. Remember back in chapter 5 (page 66), where we explained how almost no one values the carrying cost of their land? We've finally arrived at our chance to rectify this foundational error. Never forget, your local bank, main street businesses, and Wall Street all have one major thing in common: They insist upon a fair return on their investment. If not, they eventually go bankrupt. Based on American agricultural history, chances are extremely high that your farm will be no different.

From what I've observed among my farming peers, outside investing is the one economic necessity that is probably the most neglected, or perhaps even intentionally avoided. Many farmers think that the only way they can retire is to sell their land. Sadly, in most cases, this ends up being true, but only because they haven't been charging a fair price all along. Of course, you can't reasonably expect to get $25 per pound for your tomatoes, or $40 for a package of sausage, even though this might accurately reflect the true price of a fair profit, with land factored in. Let's be honest. You'd be laughed out of town if you set prices like these.

So if you're unable to pass these costs on to consumers, then you must protect your farm by strategically creating a fair return some other way. Remember the 10 percent that you've saved each week for investments? I'm convinced that investing this money *off the farm* is the best solution for generating a steady, contemporaneous return on your land expense (contemporaneous, as opposed to having to eventually sell the land in order to unlock a return). Not only does investing off the farm diversify your financial holdings but also diversification (multiple eggs in multiple baskets) offsets risk.

No doubt, investing can seem intimidating. Everyone has heard the horror story of someone who got taken to the cleaners because he bet his savings on some hot technology stock, riding it from $500 down to $5. This is a cautionary tale if ever there was one, and the reason that many people won't risk a dime outside of a savings account.

Of course, this is head-in-the-sand reasoning. Just because someone made a mistake doesn't mean that the entire system is faulty, any more than if something happens to be complicated, you can suddenly become intellectually lazy. Think of it this way: Farming is already so rife with complexity, failure, and risk, it actually makes investing look safe by comparison. There's surely a reason that the stock market has been around for more than two hundred years. So how do you take sensible advantage of some of these off-farm financial tools?

First off, never act like the guy who bet his savings. He did practically everything wrong: He made a large bet on a highly speculative equity (often, stock prices are bid up on potential, not actual profits) and didn't have a safety net to fall back on. Recall, the budgetary strategy outlined above is the precise opposite of this. Using the 10 percent rule, I make small-and-steady investments in equities that I understand and admire—always choosing ethical businesses that represent my personal values—and ones that bear a reasonable rate of return. Then, monitoring them occasionally, I hold them for a long time. It takes a tremendous amount of research and due diligence, but these opportunities do exist. And remember, as Ellen pointed out in the chapter 9, making a profit is a good thing! It gives you choices to do what you want and allows you to support the causes you value. For me, one of the things I value most is keeping my farm in business, in order to keep growing more good food.

As for the risk? Remember, by following the three tenets of the money triangle, you're already earning, saving, and giving as much as you can. Hence, inherently, you're never investing more than you can afford to lose. A safety net is naturally built into the system.

Second, and just as important, never act like those well-meaning types who are so risk-adverse that they actually lose money each year in their savings accounts. That's right, I said lose money. Sometimes, you can be so afraid of risking your money that you sleepwalk right past an opportunity. Don't fool yourself. The only time that keeping money in the bank should be viewed as an investment is if you decide to buy stock in the bank itself.

You need to understand the math here, and fortunately, it is quite straightforward. Inflation averages 3.2 percent annually, which means that one dollar today will have only have ninety-seven cents of purchasing power next year. The federal government prefers that inflation remain closer to 2 percent, but even at this lower number, the purchasing power of $50,000 is reduced to $30,477 over the course of twenty-five years (it's easy to plug this number into an online inflation calculator). How does this impact the hard-earned money sitting in your savings account? A typical bank currently offers .01 percent interest. You read that correctly: point zero one percent. That same $50,000 dollars, compounded at .01 percent interest over twenty-five years will be worth $50,125.15. That's a gain of $125.15.

So let's get this straight. If you leave $50,000 in an average savings account for twenty-five years, you'll make about $125 in interest. But in actuality, it will lose about $20,000 worth of purchasing power, because due to inflation, your dollars will no longer be worth what they were when you deposited them.

Whoa, Nelly! We're gonna need a better plan.

As farmers, there's no retirement package 401(k) plan. That is, unless you plan for one, and invest in it. So you must. It's up to you to do the hard, disciplined work of not only setting aside a portion of your budget for investing but also actually educating yourself about different investment vehicles and making steady contributions. Each week, when I take farmers' market money from my little tin cup and allocate it into envelopes, the 10 percent for outside investing isn't just helping me unlock a return on my land expenses—it's helping me to be able to make choices:

- to retire without selling the land
- to pass the land to the next generation with a sustainable balance sheet
- to possibly offer a cash inheritance, if the next generation doesn't want to farm
- to possibly hire a manager to replace me, if needed.

Remarkably, none of these options involves selling the farm! Again, having choices is a wonderful feeling. Rarely does anyone think about saving for retirement at age twenty, thirty, or sometimes even forty. But this is precisely the time when you *need* to be saving, and this is when you can build and accelerate financial protection for your farm. If you don't get excited about this idea, then you should probably check your pulse!

The following is a list of some investment vehicles that we have tried, and a brief explanation as to why. Each type has scores upon scores of books written about it, and we strongly encourage you to pursue your own education in greater depth and detail. Talk to a well-qualified financial advisor before making any investment decisions. Seeing your hard-earned money multiply is deeply satisfying.

OFF-FARM INVESTING OPTIONS

CDs

Short for certificate of deposit, a CD is a glorified savings account at a local or national bank, but one from which you can't draw money without penalty. Typically, you'd use one when you need to park your money securely for six months to a year, and draw interest that is normally many times higher than that of an average savings account. That said, these rates can swing wildly, and we think it's probably best to avoid locking up your money at a rate any lower than 4 percent. Higher rates have certainly existed in the past and will probably return again. It's a tool worth exploring, because the risk is practically zero.

Real Estate

We've already emphasized ensuring a profitable return on your farmland investment. So here, we are specifically talking about investing in properties that are located off-farm—specifically, rental houses. Buying rental homes can be a terrific way to create annual capital appreciation, while also generating supplemental monthly cash flow in the form of rent. Moreover, if you are handy, then fixer-uppers can often be purchased at a price far below move-in condition houses, and can be restored at relatively low cost. One of the first things I did with my 10 percent stash was to mortgage (after a hefty 50 percent down payment) a dumpy old house, gradually fix it up on dark winter evenings after farm work, and rent it. The appraisal immediately soared by $40,000 dollars, and I was able to ask for a monthly rent of $1,000. Within five years, I had paid the house off free and clear, and, to this day, I make 10 percent annually (including appreciation) on my investment.

Tax Liens

Back in chapter 5, we mentioned how tax liens can be used to acquire land. But they can also make for excellent low-risk investments. As a refresher, tax liens are government-auctioned rights to assume other peoples' delinquent tax bills, which you pay off on their behalf, and for which you then receive payments, plus interest—often as much as 18 percent annually. Better yet, if the taxes aren't paid after several years, ownership is transferred to you as compensation. This can be a tremendous economic windfall. I have invested in tax liens for more than a decade, and once received a property valued at $48,000 for the investment price of $2,600. Over the three-year period, that was a 164 percent annual return. Not too shabby, if you can get it!

Life Insurance

We'd be remiss not to include life insurance on this list. First of all, as we've emphasized, having peace of mind is invaluable as a farmer. A career in agriculture is statistically one of the most dangerous jobs out there, so knowing that your family and business are financially protected in the event of worst-case circumstances is vitally important (please also investigate "key man" insurance, which can be a very valuable asset, if you suffer a short-term debilitating injury such as a broken leg). But life insurance—especially *whole* plans, as opposed to *term*—can also be a great accumulator of long-term principal, which can be withdrawn in old age for your retirement. Ideally, we'll all live a long, prosperous, and healthy life. But while you're doing the meaningful work of farming, it's comforting to have one additional safety net in place, especially for a very reasonable monthly price.

Stocks

My final and favorite investment is in the stock market. Contrary to popular perception, buying stocks can be a low-risk and low-effort vehicle to accumulate wealth—that is, when done properly.

Simply put, buying a stock means that you are buying a share of a company's future profits, as well as whatever assets they presently own. Often, these assets can include cash, paid out as a dividend to shareholders (a dividend is a portion of the profits that are paid to you, the shareholder, typically four times yearly). Dividend-paying stocks are my favorite types, because not only does this typically mean the company is financially stable but, historically, dividend payments increase each year.

The takeaway here is that the dollar you invest today not only gives you a historical average rate of return of 8 percent annually but also can provide a 2 to 4 percent cash dividend on top of that. This has the potential to compound forever. If it sounds too good to be true, it's not, technically. But it *is* undoubtedly complicated at first, and warrants as much research as you can do, including hiring a good financial advisor. There are absolutely no guarantees in the stock market. My advice is to read a couple of dividend stock books and consult a professional that you trust before investing. Compounding interest is your friend. Small investments now can really add up in the long run.

Target Retirement Funds

In our opinion, though, the strongest—and perhaps easiest—recommendation is to purchase a target retirement fund, offered by many different companies. These are highly diversified, multi-stock (mutual fund) vehicles that automatically recalibrate into other less aggressive positions (such as bonds and cash) as you get older. You can usually set these up with automatic monthly withdrawals from your bank account. Best of all, up to a certain point, contributions can be deducted from your annual income.

Ultimately, your goal as a new farmer isn't just to take care of the land, it's also to generate a fair return for your work. Small contributions, and I'm talking as little as five dollars daily here, really add up. But if you never start, you miss out on years, decades, or perhaps even a lifetime of accumulation of compounding interest. Create your safety net so that you—the caregiver who grows food for your community—can do a better job of taking care of others.

And when your work is finally done? You want the option to retire comfortably. If you can accomplish this without selling your land—the heart and soul of most farmers—then all the better! If you plan in advance, these are extremely reasonable goals. Don't let anyone ever convince you otherwise.

Let's farm with our hearts, invest with our heads, and retire rich.

CHAPTER REVIEW QUESTIONS

1. Who is the most successful investor that you know personally? Take them out to lunch, and listen.

2. Have you ever created a budget for your personal finances? If so, what strategy did you use, and was it successful? Did you ever stray from this budget? If so, what were the consequences?

3. Consider friends or family members who are retired. Where does their money for living expenses come from? How have they managed these resources to provide for a stable, long-term retirement?

CHAPTER ELEVEN

THE THREE GOLDEN RULES OF FARMING

Forrest

In the United States, unless you start your farm in the middle of a remote, primeval forest, there's a 100 percent chance that some or all of the soil has been eroded, demineralized, mined of nutrients, or even blown away. In 1935, wind-borne dust from Oklahoma reached all the way to Boston, New York, and Washington, DC, dimming the midday sky for five hours. With their suits and ties covered in dirt from a 1,200-mile distance, congressmen finally settled their partisan bickering and funded long-overdue soil-conservation programs. To this day, centuries of plowing, deforestation, and nonstop harvesting have left hundreds of millions of acres with a fraction of their original fertility.

Factor in lack of water retention, increased salinity levels, soils devoid of organic matter, compaction from heavy machinery, and a reliance on commercial fertilizer that borders on being mandatory for conventional systems, and the picture can seem pretty grim. Our soil was once our heritage, as well as our future. Now the next generation of farmers has been left to fix all of these problems—one farm, one harvest, one acre at a time. It's a national crisis unfolding all around us, unquestionably one of the greatest issues of our era.

I faced these same daunting odds myself when I first started. I soon discovered that, as on most farms in the Shenandoah Valley, much of my soil's original fertility was already at the bottom of the Chesapeake Bay, washed away by centuries of tillage, lack of cover cropping, and livestock allowed to wallow for hours in our spring-fed creek. Our soils were so devoid of water-absorbing organic matter that, on some parts of the farm, it was debatable whether we were growing pasture or red clay. On hot summer days, I could see moisture evaporating straight out of the ground, opalescent, a shimmering mirage with the Blue Ridge as a backdrop.

But it wasn't just the soil that was in rough shape. Our two-hundred-year-old barn was so dilapidated that it soon fell over in a strong wind; we propped it back up, but another storm came through and knocked it down again. A fleet of ancient tractors was constantly broken down, and our equipment was so worn out that the steel was as thin as tissue paper and an accident waiting to happen. Everywhere I turned, every time I tried to do some work, something else was broken. From the mid-1970s to the early 2000s, though we tried our hardest, our family farm never made a nickel of profit and was, instead, subsidized through the off-farm jobs of my parents, federal crop assistance, and my own volunteer labor.

Median farm income, median off-farm income, and median total income of farm operator households, 2014-2017F

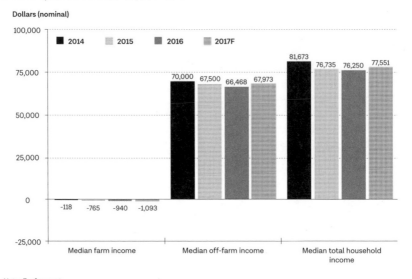

Dollars (nominal)

Note: F = forecast.
Source: USDA, Economic Research Service and Natonal Agricultural Statistics Service, Agricultural Resouce Management Survey. Forecast as of November 29, 2017.

Fast-forward almost two decades, and, while my own farm has made a tremendous turnaround through the adoption of sustainable practices and direct marketing of products, the national statistics haven't improved. The median American farm (see chart 176) still depends on off-farm income to stay afloat. Costs of machinery and infrastructure far outpace commodity price growth. And most farmers remain shackled to a system reliant on chemical inputs, practices that frequently accelerate the depletion of an already depleted soil. So what's to be done?

I once heard someone say, "I'm never so impressed by what a person has, as what they *do* with what they have." For me, this cuts to the core of agricultural success. There's nothing magical about my farm, no special advantages or secrets. For thirty years, our place was a broken-down, unprofitable money pit, plain and simple. Multiply my story by two million, and you have some sense of the last fifty years of agriculture. This is our American farming landscape, our soil, our mutual inheritance. It's also, undoubtedly, a clarion call to action.

The only way I know how to solve these problems is through the most common magic of all: waking up each morning for several decades in a row, sticking to a handful of no-nonsense rules, and heading out to work.

It's time once again to take stock of your assets. Regardless of the liabilities your farm may come with, remember that you've also been given incredible gifts and tools to help you repair a broken system. These include:

- free resources such as sunshine, rain, nitrogen, and carbon, endlessly supplied by Mother Nature
- knowledge that sustainable practices not only work, but work well
- intellect to solve the challenges that will inevitably slow you down
- passion to give you energy and maintain your spirits
- like-minded community to provide support, encouragement, and problem-solving resources

These are all freely available, awaiting your conscientious management. So what are you going to do with what you have been given? If you've read this far, then you already possess the passion required to start a farm—the internal drive and the emotional resiliency. Nothing's going to come easy, but the answers are out there. In fact, most of them are waiting for you in plain sight. You already have the tools to fix up your farm.

Now, far more than luck, you need some useful rules to help beat the odds. It just so happens, I keep a few handy.

RULE 1: IF IT'S BROKEN, STOP AND FIX IT

On my farm, there are probably fifteen thousand fence posts, each holding up four strands of electrified high-tensile wire, stretching for miles upon miles around emerald pastures thick with clover, sheep, and cattle. The fences keep our livestock where they are supposed to be—namely, contentedly grazing in their meadows, and away from the busy four-lane highway less than a mile from our property's edge. I'd never presume to know what any of my animals are really thinking, but I suspect, however, that they are far happier munching bluegrass than playing bovine dodgeball with tractor trailers along US 340.

Hence, fence. Over the years, using profits from farmers' markets, I've overhauled the entire infrastructure, replacing the rusted wire and hand-split locust my grandfather installed sixty years prior. Each new post cost me eight dollars, and another two dollars of labor to place it two-and-a-half feet in the ground. The electric charger, powering dozens of miles of perimeter and cross-fencing (fencing that, incidentally, is roughly one dollar per foot to install), costs one thousand dollars. Factor in the watering systems, the price of the cattle, and farmhand salaries, and it's obvious that there's a lot of money on the line.

Yet, despite these sizable numbers, the entire system hinges upon a handful of two-cent fence staples, tightly securing the wire to the post. Lose one tiny U-shaped piece of steel, and suddenly a strand springs free. Lose two, and a steer will stretch his neck through the fence, munching the proverbial greener grass on the other side. Doing so, soon the third staple pops loose, then the fourth . . . then the steer itself. Not to be left behind, my entire herd will quickly follow the leader. Where do they go from there? Oh, right. Off to play dodge-cow on the highway.

Benjamin Franklin famously observed, "For want of a nail, a shoe was lost." On our farm, the stakes are significantly higher than footwear: For want of a staple, an entire herd can be lost! Now, if you ever need a good jolt to clear out your arteries, I highly recommend receiving a call from the local sheriff at 2:00 AM, gruffly informing you that your cattle are loose on the median. I've responded to this call three times over the course of my career, and I'm honestly not sure whether the next call will cure me or kill me.

But back to that missing staple. In this instance, the failure wasn't the fault of the system. It was the fault of the operator. It's easy to spot a broken post or wire lying on the ground, but how easy is it to notice a missing staple? And, even if it's spotted, you must have the discipline to interrupt whatever you're doing for thirty seconds, and fix it—that is, presuming you had the foresight to keep a hammer and extra staples handy. What seems like a minor inconvenience can lead to a major headache, if left unattended. Because I understand the potential ramifications, stopping and fixing the problem has always been worth a two-cent staple and thirty seconds of delay to me.

It could be a missing staple, a tire that's nearly flat, a tractor with low oil, torn drip tape, a breach in the deer netting, a frayed electrical wire in the barn, or a power saw that needs its blade tightened. On a farm, when things are in need of repair, the consequences can range from expensive to dangerous, to even deadly. None of these options are recipes for a good day.

This rule goes much deeper than tacking down a loose barn board or rehanging a gate. Remember, the wording is "If it's broken, *stop* and fix it," not "Hey, get around to fixing that broken thing sometime, if you happen to remember it." Allow me to politely explain that, on a working farm, the day never arrives when we suddenly have extra time. Life always gets in the way—seeds must be planted, fruit harvested, pigs loaded, and cows milked. I

SIMPLE IS BEAUTIFUL

"If it's broken, stop and fix it" teaches us that simple systems are often the best, especially for a lone operator. No matter how conscientious you might be, things will always break, and accidents will happen through acts of nature. Often, the simpler your systems are, the easier they are to check, fix, and improve. Ease of use is also invaluable when training a new staffer or a part-time helper, so you can take a day off. It's far easier to train someone to drive a staple than to ask them to build a new fence.

once considered renaming my farm Round Tuit Acres, because growing up, we always used to say, "Oh, we'll get around to it, eventually." I'm still waiting for a *round tuit* to appear.

To be sure, the word "stop" is an intentional action verb in this rule. It really means just that: stop, you busy, distracted human! When you pause, you not only take time to make the repair but also you get the opportunity to assess and reflect. Perhaps there was a reason something broke; maybe a simple tweak could improve the system for the future. Maybe it's clear that something is completely worn out, now that you've taken the time to stop and fully examine it. Pausing and assessing almost always yields positive results.

This line of reasoning works on practically all levels:

Marketing: When your products aren't selling, your marketing or sales channels need attention. There's only so long that your sales can diminish before you're faced with a cash-flow crisis. Take the time to stop and fix it. Assess your product line or sales channels and make changes, or take the afternoon to investigate how other producers are marketing their products. Often, the solution is as simple as a ten-degree pivot, not a full about-face.

Communication: When you've argued with someone in your family or a business partner, and feelings become ruffled, don't walk away in a huff, or, worse yet, simply ignore it. Take the time to stop and fix it. Everyone wants to be heard and respected, just as they want a chance to participate and lend creativity. Bad relationships can ruin a farm faster than bad soil can.

Well-being: If you find yourself going seven days a week for weeks on end, working sixteen-hour days just to get the job done, then, for your own sake, you need to *stop*. Then—after you've caught your breath—you need to *fix* it. No one is meant to work seven days a week, no matter how robust your constitution. If your productivity doesn't suffer first, then your health eventually will. Burnout is real—both physical and emotional. One of the best decisions I ever made was to budget for a part-time helper and reclaim a day off.

RULE 2: PUT IT BACK WHERE YOU FOUND IT

For me, this second rule came about for a very practical reason: to keep me from going completely nuts. I grew up on a farm where, frequently, no one knew where the sole working tractor had been parked. Of course, we knew it had to be *somewhere*; after all, it weighed several tons and was bright blue. And we knew it hadn't been stolen, because this would require having the key—a key that was invariably left in someone's jeans, casually tossed onto the workbench, or dumped onto the dresser mixed with pocket change and bent nails. After a typical hour of searching, the key might be discovered, then after another hour, the tractor. It only required budgeting two hours in advance. Then, after walking a mile to wherever the tractor had been left, we could reliably get our day started by the crack of noon.

When I took over the farm in my early twenties, I realized that this issue extended far beyond the tractor. I discovered tools that had been forgotten along fencerows, now rusted and nearly useless. Extra lumber from construction projects was left outside, and quickly rotted. The workshop—the de facto hub of the farm—was so disorganized and crammed with all sorts of random, broken junk, scrap, boxes, buckets, and trash that there was no place for anything else.

Our disheveled farm needed one of those extreme-makeover shows, where professional OCD whirling dervishes appear in an oversize van, work their frenetic magic, and leave a picture-perfect farm in their wake. Fingers crossed, I kept looking down the lane. No van ever appeared.

Instead, I did the next best thing: slowly but fastidiously put things back where they belonged. One tool at a time—one screwdriver, one pail of nails, one set of wire strippers, one jack, one oil pan—the farm gradually began to transform. When I needed an extension cord, for the first time in my life, I knew exactly where it was. When a hose needed repair, I simply went to the box where I kept spare parts. If I needed some WD-40, the can was right where it was supposed to be. Unlike on the farm of my youth, my $^9/_{16}$ wrench never grew legs, jumped off the hook, and walked away in the middle of the night. Instead, each morning, I found it reliably hanging right where I left it, between the $^1/_2$ and the $^5/_8$. What a delight!

The put-it-back-where-you-found-it rule worked miracles for me. My efficiency soared, because I wasn't running in circles looking for tools or parts. My costs decreased, because I wasn't driving to town to buy a replacement posthole digger that had been lost God knows where.

Aesthetically, the farm became more beautiful, too. For decades, trash and junk piles had been allowed to accumulate on the farm, agricultural garbage (old pesticide bottles, broken-down equipment, empty oil cans) left to lie practically wherever it fell. I soon adopted a corollary to my rule: "Trash attracts trash." If I didn't put things back, disorganization soon reigned. A flat tire, discarded in the corner of the barnyard, mysteriously attracted another tire, then a third. Turn around twice, and suddenly there was a tire pile! Mangled balls of rusted fence wire seemed to increase in circumference overnight, and soon attracted protective peripheries of trees. The most diabolical of all were dead tractor and truck batteries, multiplying like rabbits every time I looked away. It wasn't enough to simply put things back where I found them—I had to remain on vigilant trash patrol. The garbage was clearly conspiring to get me.

Eventually, just as I had with rule 1, I realized that there was more depth to rule 2 than I first imagined. As a livestock farm, we had always made our own hay, firing up our worn-out equipment summer after summer, and mowing down swaths of grass to be bundled and trundled away until the winter. But after a decade of making hay myself, I could clearly see that our yields were decreasing, and, worse yet, undesirable plants such as broom sedge and thistles were springing up in our rapidly thinning fields. Something wasn't right.

Eventually, it struck me: We weren't putting things back where we found them. Each year, we'd harvest hay in one field, then feed it out on another. In essence, we were robbing our fields of nutrients, shuffling potassium and phosphorus around the farm with ever-decreasing returns. The soils were already depleted, and now we were stretching them ever thinner, never fertilizing them in a way that would allow them to fully flourish. Instead of putting it back, we were constantly taking it away. The results were evident in our paltry yields.

If I was ever going to build my soils back to their historic potential, I needed to stop taking and start giving. I took what I considered a bold leap at the time, and stopped making hay entirely, purchasing it from other farms instead. It wasn't enough to put back just some of the nutrients; we needed to be putting lots of it back. To accelerate this, I simultaneously ramped up our pastured hog and poultry operation, purchasing grain to increase the amount of manure that would fertilize our fields. At the same time, I fed mineral-rich kelp, packed with seaborne micronutrients, which were passed along through the manure and urine of our animals. All of this, along with feeding (not making) the hay, turbo-charged our inputs. At long last, we were finally putting it back where we found it—or at least, where it used to be.

On the most deficient parts of the farm, the land responded almost immediately. Everywhere the manure landed, within weeks, the grass grew three times as thick, and grew twice as high as the surrounding pasture. Our soils were so starved for nutrients, they responded like a dying patient receiving medicine. What seemed like a risky financial move at the time—halting our hay production after having invested so much into the equipment to do it—ended up paying us back in a few short years, through increased yields. Achieving full-farm results required more time, but, within a decade, soil tests showed nutrients well on their way back to robust levels, and our organic matter had tripled, from 1.8 percent to 5.5 percent. When we finally committed to "putting it back where we found it," the investment paid huge dividends.

Having strong soil fertility is akin to having money in the bank. When you have enough cash in your bank account, you can occasionally draw a little interest off of it (for the sake of this analogy, recall when savings account rates used to be higher than inflation!). But if you can build your account to a very high level of principal, at a certain point, money starts making its own money. It's no stretch to understand that if you keep putting the money back, allowing the numbers to compound, there eventually comes a point where you can live off the interest without ever touching the balance. This happens all the time in finance.

So, too, with your soil. When you have nutrient-rich soil, it is actively building humus, fostering nitrogen and sugar exchanges, sequestering carbon, and retaining more moisture. Eventually, when your fields are healthy enough, you can reach a point where your profits from the yields (food, fiber, feed, or flowers) will far outstrip the modest fertility amendments needed to maintain their performance. Of course, nothing comes from nothing, and you'll always need to fertilize and "put it back" as necessary. But your soil has the amazing ability to generate "interest," once it is fully remineralized. This requires a balance of human management, time, and the enhancement of natural mechanisms.

It's important to understand why the philosophy of "putting it back where you found it" usually runs counter to the way agribusinesses—the corporations that provide chemicals, machines, and services to farmers—typically try to solve agricultural problems. Recall how this chapter started, with the fact that nearly every farmer must contend with depleted soils. In an attempt to fix this on a large scale, commercial fertilizers are typically used, in conjunction with high-yielding GMO seeds and their accompanying herbicides. Short-term, these products seem to offer a miracle solution, supercharging soils incapable of producing such harvests on their own. It's like dumping a tanker truck of 5-Hour Energy onto your fields.

But over time, soil tests reveal that this is a system of diminishing returns. By constantly treating the symptom (low yields), you never truly addresses the actual problem (soils depleted of nutrients and organic matter). Instead, you become reliant on these costly solutions, and are eventually unable to achieve the same yields without more fertilizer and the latest variety of GMO seeds. The chemicals might help grow the crops, and the engineered seeds some relative abundance at first, but the soils themselves grow weaker and more depleted over time. This is the opposite of building up your account and drawing interest. Instead, it's like you paying the bank a monthly fee on top of the privilege of giving them your money.

It's long been scientifically demonstrated by major research institutes and educators, such as Iowa State University (Ames, Iowa) and The Rodale Institute (Kutztown, Pennsylvania), that sustainable farming practices offer a proven track to soil restoration and naturally higher yields.

So why don't more farmers use these sustainable methods? Because putting it back where we found it takes *time*—as much time as we can give it. Plainly stated, hardly anyone in modern agriculture has the luxury of time. The system doesn't allow for it:

- The bank won't let you skip a payment while you heal the soil.
- The tractor dealership won't misplace your invoice while your compost matures.
- The fertilizer, seed, and services companies won't suddenly become philanthropists while you're building enough soil carbon to ride out the next drought.

Everyone wants to be paid, and so trying to squeeze the remaining few dollars out of their bankrupt soil—yet with commodity prices always calibrated to offer the lowest price possible—few conventional farmers have the luxury of looking more than a year or two down the road, much less the five years required to significantly start boosting yields from sustainable practices.

A red-dirt cornfield is a pretty bleak sight, but with the right management, it can be naturally transformed and fully rejuvenated. R. Buckminster Fuller once said, "There is nothing in a caterpillar that tells you it's going to be a butterfly." The same could be said of soils. They must be given time to mature and emerge from their chrysalis.

This, by far, is the biggest lament I hear from conventional farmers: They don't have the time to give their farms what the soil truly needs, those precious years required to "put it back where they found it." Yes, they believe in the merits of sustainable systems, even if they might not personally use them. But regardless of what they'd prefer to do, they are locked into a system that prohibits anything but staying the course.

If we're ever going to shift this paradigm, then as new, sustainable farmers, we have to stick to the rules from the outset. We have to stop taking, and start putting it back where we found it. The soils are bankrupt. If it's broken, whether the problem is big or small, we have to stop and fix it. And finally—perhaps most important—we need to do what we say we're going to do. Fortunately, there is a rule covering this territory, as well.

RULE 3: DO WHAT YOU SAY YOU'RE GOING TO DO

Rules 1 and 2 produce fabulous results, but only if we actually perform them. Isn't it funny how life works like that? We're all planners and dreamers in our own ways—this is part of being human, and an important trait of being a farmer. There comes a time, however, when we move beyond daydreaming and into a world of tangible results. When it comes to farming, if we don't do what we say we're going to do, then we face real consequences. A day quickly arrives when the only BS on the farm should be out where it belongs, fertilizing the pasture.

As usual, this rule has big ramifications. Sure, you need to actually stop and hammer those missing staples—just like you say you'll do. And it's important to put that hammer back where you found it—again, as you say you will. There are solid, pragmatic reasons for you to stick to your stated principles. After all, if you can't count on yourself, who can you count on?

But our final rule extends well beyond day-to-day chores. When you keep your word, it impacts your relationships with your neighbors, your family, and your community. Words matter, and when you keep your promises, it fosters trust. If you earnestly do what you say you're going to do, positive ripples reach far and wide.

Take your local agricultural communities, for example. As a new farmer starting out, you'll want to introduce yourself to and become friends with as many established producers as possible, either through attending local events, joining an online group, or actually driving down the lane and shaking hands. From my experience, most farmers are very willing to share knowledge and advice, once they warm up to you.

But these first few years of relationship-building are critical for your enduring reputation. If you borrow something, return it on time and in as good or better shape than you borrowed it. If you purchase something from another producer—straw, machinery, livestock, seed—pay your bill either promptly or on the spot. If you promise that you're going to be somewhere or do something for them, be there or do it. Trust is golden in the farming community, and it's also a two-way street. Farmers might be forgiving, but they rarely forget. Drop the ball a few times, and word will quickly get around. Reputations stick.

This rule is crucial for your personal relationships, as well. Speaking from experience, I've squandered plenty of goodwill over the years by being chronically late—late for an appointment, late for a meeting—even late for dinner or for a date. On the farm, things always seem to take twice as much time as I think they will. Maybe the cattle were pokey that day, or I spotted a leaky hydrant from across the field, or the fan belt on the truck suddenly gave up the ghost. On any given day, there could be a hundred legitimate reasons why time gets away. But in most instances, it's just my own stupid fault, plain and simple.

Here's a news flash: Other people have schedules, too, with bona fide time constraints and challenges of their own. You need to honor this. Your farm mustn't regulate your life to the point where you constantly drop the ball with your partner, your staff, or your children. Sure, the farm is important. But unless you've already made up your mind to be a hermit, you'd better come in for dinner when you say you will—or you might find yourself living *la vida solo*. Time, ultimately, is all you have. Pick some flowers for your sweetie, be a horsey for your kid, and celebrate that wonderful food you grow with the people you love.

Finally, consider your community at large. Over the years, the public perception of farmers has evolved from bib-overalled bumpkins to figures worthy of esteem and admiration. Few occupations today are as respected as farming. We're in a position of trust in our society, so let's not squander it by letting folks down. Do what you say you're going to do.

This means that if you're a vegetable grower in Kansas, and you sell at a producers-only farmers' market, don't throw on a few extra boxes of wholesale tomatoes grown in southern Texas. It's not only cheating, it violates consumer trust. If you're a free-range livestock farmer, don't suddenly pen your chickens into a barn for six months out of the year, just because it's colder outside. That's not the product you advertise, or the eggs people think they're buying. If you're an organic berry farmer, don't suddenly spray your brambles with pesticide, simply because the stinkbugs are particularly bad this year. You've made a promise to grow a certain way. Now stick to it.

Trust is fragile, especially among consumers. And once it's lost, it's very difficult to regain. But trust isn't just isolated to one farm. Having blogged over the years about the merits of farmers' markets, CSAs, and roadside stands, I can't begin to count the number of comments nationwide from consumers who say they've been misled or outright cheated by producers who claim to do one thing, then do the opposite. When a farmer doesn't do what he says he will, it makes all of us look bad. We owe it not just to our communities to maintain high standards, but to our fellow hardworking peers.

And don't just keep your word for others. Do it for yourself. Be the farmer *you* want to be. Farming is already challenging enough without compounding it with self-doubt, guilt, or shame. Stick to your principles—improve them and expand them—and sleep soundly at night. This is a worthy path. "Do what you say you're going to do" seems simple enough on the surface, but it's the execution that's hard. Repetition helps. So does the power of conviction.

These are darn good rules, but I'd never claim that they're easy. They will pester you, confound you, and relentlessly buzz in your ears until they stick. Then, once you're thoroughly exasperated, and have learned from experience that these rules truly work wonders, you get to *enjoy* them. By applying these principles, you put your farm on solid footing from the start. Systems, soils, and even your relationships and personal well-being will all benefit.

No doubt, you're bound to come up with some great rules of your own, wisdom and advice that will be invaluable to others. Pay it forward.

CHAPTER REVIEW QUESTIONS

1. Sometimes you're unable to fix certain things. For instance, the electricity might go out for an entire day or the well pump might fail, leaving you without water. Imagining your farm, what steps might you take to prevent or adapt to such a situation?

2. In order to "put it back where you found it," you'll need to understand your soil's specific chemical needs. Have you gathered soil samples before, or read a soil report? If not, how will you go about learning these foundational skills?

3. Think of a resolution you've made, one intended to change your life for the better. Were you able to stick to what you promised? What was the end result?

· CHAPTER TWELVE ·

LOVE, WORK, AND HARMONY

Ellen

Y ou may be surprised to learn that, while much farm work is done alone, where a person can be as grumpy or dour as they like with little consequence, succeeding in this business depends on maintaining excellent relationships. That's the thing about business, right? It's done with other people. So, if you don't have good relationship skills, you have three main choices: Choose another career, have a great business partner who covers for you, or develop some new relationship muscles. We endorse the third option.

In case you haven't heard this enough already on TV or in your reading, it's all about communication. *How* you communicate, both tone and content, matters. *Why* you communicate (what is your agenda? are you being honest or manipulative?) is important. *When* you communicate makes a huge difference. Are you holding onto misunderstandings and snapping in the heat of the moment, or are you taking time to cool off, reflect, and respond when the timing is more appropriate?

With so many unknown variables to contend with in this business, it's reassuring that good communication skills are easily acquired with practice. You must also cultivate the skills to communicate with many different kinds of people, from employees, service providers, and peers to family and other members of your community. In the farming business, there are multiple arenas in which the quality of your relationships will help shape your success or failure.

LABOR

Go to any farming conference or hang out with any group of growers, and, typically, within a matter of minutes, there will be whining and complaining about labor. First off, the old saying is mostly true: Good help is hard to find (though, as we discussed in chapter 2, great leadership is even *harder* to find—and great leaders have a way of attracting and retaining good help). Second, especially for new farmers, it's often more difficult than expected to properly train people. And third, once they are trained, it can be hard to keep good folks around.

Want to make an annual income of $40,000 to $50,000 in farming? Sounds good! However, if you plan to do this alone, a reality check is in order. It is very difficult for a single human to grow, harvest, and market $125,000 worth of any agricultural product without help. (That's about the total sales you'd need to generate that salary.) You may be lucky enough to have children or relatives who can work unsalaried, but this comes with its own risks. Though this might sound good for your bottom line, expecting family members to work without pay can be even more complicated emotionally than managing non-relative employees. Remember, this chapter is about building relationships, not sabotaging them. We can cite multiple examples of farms being torn apart due to the frustrations and resentments of family members who are asked to work without appropriate compensation. Suffice it to say, almost all farms need to hire help, whether there are family members or not.

Workforce Relationship Models

It's imperative to understand the various relationship models that are available to farmers and their help. Each have their pros and cons, and are never as simple as they appear at first glance.

Fortunately, there are several management rules that apply to all the labor models to be discussed below. Foremost, remember that no one will work as long, hard, or as well as you. It's your farm, not theirs, so have reasonable expectations of your labor force.

- Set clear, consistent, and attainable goals.
- Give and receive feedback generously. It is up to you to create safe and open channels of communication.

- Don't publicly chastise anyone.
- Be likable, humane, and express interest in others' lives.
- Don't gossip about your team members.

Employer and Employee

Like in most businesses, the employer-employee model is the standard operating procedure for many farms. The process is fairly straightforward: Run a help-wanted ad, interview promising candidates, check their multiple references (always do this, and thoroughly; if your intuition alarm ever goes off, pay attention to it), then make your hire. But really, this is just the start. It also means that you will have various legal responsibilities, and that can scare some farmers away from ever hiring outside help. If you wish to create an income beyond what you—as a solo farmer—can grow, then this is something you must get over. You will learn how to:

- withhold and pay federal income taxes
- withhold and pay employer share (7.5 percent) of FICA for social security and Medicare
- withhold and pay state income taxes
- pay federal unemployment tax
- pay state unemployment tax
- get workers compensation insurance

The last item on this list is something that many growers avoid purchasing, and that's a mistake, in our opinion. Worker compensation insurance helps protect your assets from being sued away from you in case an accident on your farm injures or (God forbid) kills an employee. Of course, those kinds of events are what you try hard to avoid, but accidents do happen sometimes.

The strength of the employer-employee model is the clarity of the power structure. You are the boss, and they are the workers. You are paying them by the hour, or by salary, for their time and efforts. It is up to you whether to include other forms of compensation—such as food, housing, or health insurance. There are tax consequences to some of those benefits, and you should have a clear understanding with your accountant of how to handle them. In

INSURANCE PROTECTS A DEAR FRIENDSHIP

Once upon a time, I agreed to hire the daughter of a dear friend for a week. The goal was to show this kid how hard farm work was, before she set off on a farm-hopping international trip as a WWOOFer (World Wide Opportunities on Organic Farms). Midweek, while working with the crew to get T-posts up for a tomato trellis, this young woman bonked herself on the head with the post-pounder. She fell to the ground, and then left for the rest of the day.

Her doctor diagnosed a concussion. Her symptoms did not improve for a long time—she had vertigo and light sensitivity for months. Our insurance company paid all her bills (roughly $25,000) and finally offered her a settlement of $100,000 to fulfill its responsibility for the injury. She took the settlement, and, fortunately, she did fully recover.

The moral of the story is twofold. First, we did not have to pay that $125,000 out of pocket. The young woman had all the medical care she needed, without me having to do anything. Second, I remained friends with her mother, because it wasn't my negligence that caused the accident and the insurance handled the whole thing. This is why you get insurance, because weird things happen. (P.S. We subsequently implemented a hard-hat policy for anyone working with a post-pounder.)

our opinion, it is simpler to keep these transactions separate.

For example, though you may have housing on your farm, it's far better to pay employees a higher wage and have them subsequently pay you rent for housing, then it is to pay a lower wage and have this included within a compensation package. It may seem silly to do multiple transactions, but it helps make clear the value of both the labor and the housing. Also, in the event that someone needs to move off the farm but stay employed with you (this can happen for many reasons: marriage, expanding families, spouse with a longer commute), it eliminates this potential complication before it ever arises.

Internships

As we mentioned in chapter 4, an internship is when an experienced practitioner takes on a person who wants some work experience, with the understanding that the position is temporary and will not necessarily lead to a permanent job. Applied to an agricultural setting, the farmer provides a formal educational program, including readings, discussions, lectures, and field trips. The farmer encourages and coaches the intern to perform complex tasks as an educational experience, instead of handing them off to a seasoned employee. In exchange, the learner provides some amount of manual labor, sometimes paid, but more often than not, for free.

We've often seen this model abused, particularly in small-scale agriculture. New farmers who don't have the budget to hire employees as described above (page 192) sometimes take on interns instead. In our opinion, an internship has legal and moral implications. Simply allowing someone the opportunity— and oftentimes privilege—of working for you does not qualify as an internship. Pay is typically not the sticking point; rather, it's all about the quality of the educational experience. Is the teacher providing adequate instruction to the student? Are there appropriate off-farm experiences to help accentuate the learning process? These are the types of questions that create a meaningful experience for both farmer and intern alike.

Providing an internship component is quite a serious undertaking—taking time away from the regular work of the farm in order to provide education beyond the instructions to get the job done. By definition, as a beginning farmer, you are not qualified to offer internships. But it is certainly something to aspire to, should that teaching role excite you.

Volunteers
Some farms find a role for volunteers to play. For some farmers, it may be easy to find willing—even enthusiastic—volunteers. The real question becomes whether the work done by volunteers is worth the time it takes to manage their work. Untrained volunteers may work cheerfully, but the quality of the work may be insufficient, or even lead to bigger, unexpected problems. There has to be a greater social mission to make this transaction satisfying.

CSA farms are the most likely to use volunteer labor. It is part of the temperament and mission of the CSA model to use the labor of customer volunteers to strengthen the bond between farmer and eater. For example, Vermont Valley Community Farm in Wisconsin has a force of volunteers each week who pack hundreds of CSA shares for delivery into Madison. The work is fairly simple, and the volunteers are also shareholders, so they have a vested interest in doing good work, making sure each CSA customer gets the quantity and quality of produce they deserve. Another CSA farm in Alabama has two of its members come out to the farm each week to help wash produce and pack the share boxes. Those volunteers then deliver the shares to town thirty minutes away.

If there is anyone working on the farm other than you, you must have adequate insurance, in case of accidents. Additionally, if you have any member of the public on your farm, you will need liability coverage. The details of these kinds of insurance vary by state, so work with your insurance agent to get it right. Despite your best intentions, it's only sensible to protect your assets from a possible suit in case of an accident.

Volunteers come to the farm for an enjoyable experience and to develop a sense of connection to the farm. But for most farms, no matter the size, volunteers simply require too much supervision to be very valuable. From our experience, they require lots of explanation and demonstrations, and are generally incapable of much truly hard labor, especially in trying weather conditions, for any significant length of time. Working with volunteers requires a different attitude and skill set entirely. If this is part of your core mission statement, however, it can be a mutually fulfilling arrangement.

SERVICE PROVIDERS

As you know by now, farms are complicated organisms, with many moving parts. Lots of things (and beings) need to be taken care of. Unless you are an incredible jack-of-all-trades, you will need some professional help to keep your farm running smoothly. Here's a short list of the kinds of service providers you may need in your phone contact list:

- tractor parts and service
- well service and plumbing
- vehicle maintenance and repair
- electrician
- welder
- veterinarian
- feed and seed supplier

- fuel supplier
- trapper
- fertilizer dealer
- carpenter/handyman
- refrigeration specialist

Did we forget "etc."? So here's what you are aiming for: First, whenever you call one of these folks, you want them to react favorably to your name appearing on the caller ID, so they will answer the call. Second, upon answering or returning your call, you want them to help you get what you want in a timely and pleasant fashion. This sounds simple enough, but know this: These service providers are dealing with many of the same variables as you are, plus travel distances, odd hours, employees of their own, and disgruntled customers on top of it all. To them, you are just another person with another problem, another voice on an endlessly ringing phone.

Still, you need to break through this noise, and somehow make each of these relationships resilient. As previously mentioned in chapter 4, if you do this properly and make a meaningful connection, in many cases, that person can give you free advice or help you fix the problem yourself—in essence, become your coach. This is an incredibly valuable relationship to have.

How can you help the odds of this occurring? Become a pleasant and professional customer. Enjoy your interactions with each of these people. Be patient. Ask how their business is doing, and give them referrals. Be courteous, thankful, and gracious. Be flexible when you can be, and push only when you have to. Most important, pay them on time. Nothing speaks quite so effectively in your agricultural communities as a prompt payment, and a check that doesn't bounce!

This courtship takes some time and intention. But having a small army of service providers who will pick up the phone for a free sixty-second consultation is almost invaluable. Not only will this level of relationship end up saving you time and money, the assurance that someone reliable is waiting in the wings to help will make your days more relaxed and joyful.

OTHER FARMERS

Depending on where you live, farmers might be a fairly rare breed; there are a lot fewer of us now than there have ever been in history. So it behooves us to stick together as best we can. When farms and suburbs rub against each other, some friction may result. You need friends on both sides of these issues—farmer friends and customer friends. So remember what you have in common with any farmer, whether they are growing a thousand acres of GMO corn or milking six hundred head of Holsteins. We all need access to clean air and water, sunshine, and safe ways to get our products to market.

I have been blessed to be part of a very generous community of local growers. We order supplies together. We have winter meetings to talk about successes, failures, and dreams. I lend my neighbor my root digger, and he lends me his post-hole digger and subsoiler. I keep extra plants around the greenhouse in the spring, in case one of my farming comrades has some kind of disaster and needs plants fast, as others have done for me. We hold our own neighborhood Farm Olympics every four years, where events include hay bale tosses, squash throwing, and reverse tractor driving with a trailer in tow. All this fun occurs despite the fact that most of us are in direct competition at markets.

We have found, over the years, that sharing our knowledge, tips, and assistance with other growers is rewarded with help and advice when we are the ones who need it. It may not be a one-to-one ratio or a direct reciprocation. It's more of a network of goodwill, an inherent resiliency built into our local agriculture system. Additionally, we have benefited by being able to supply our vegetable CSA and roadside stands with the good food grown by others in our community, such as meats, dairy products, and fruits.

I am told that this kind of generosity is more rare in the conventional agriculture community. In a very competitive market, the incentive is often perceived as either get big or get out—and to get big, someone has to fail and sell their farm. That results in a pervasive feeling of "not enough"—not enough markets, not enough customers, not enough land, not enough success to go around. Therefore, folks keep tips and new insights to themselves to maintain a competitive edge. This is a mindset of scarcity, rather than of abundance.

Using your tai chi economic skills, set the example of generosity and cooperation in your community. It's worth remembering that each relationship within any given community is unique, and you'll want to remind yourself to recalibrate preconceived notions. Often, when it comes to other farmers, it's best to leverage what you have in common, rather than focus on where you differ.

FAMILY AND PERSONAL RELATIONSHIPS

Farming can be, and often is, a completely engrossing profession. You know the term *workaholic*—a person who compulsively works hard and long hours. It's an expression that our culture has widely adopted, and, for many, being dubbed a workaholic is a bad thing. For good or ill, many farmers are workaholics, by definition. If a farmer, and especially a beginning farmer, doesn't work long, hard hours, then they are less likely to succeed. But it's not just farming success that's on the line. Being a workaholic can affect one's personal life, as well.

Unlike many jobs, where the work is left behind when the office closes at five, a farmer's work is often completely entangled with their personal life. First off, most farmers live on their farms (though we also know a number of growers who have either never lived on their farms or have actually moved from the land to town, and love it!). It's common for farm offices to reside within the family kitchen. Seeds are often started inside the house. Trucks, golf carts, tractors, and tools are typically waiting just outside the door, ever-present reminders of work. Livestock is usually within sight, or at least within earshot.

Adding to your likely workaholic inclinations, the weather, or your market will often require working at odd times of day. You might be up at 4 o'clock on Saturday morning for market, bunching flowers late on Friday night, or helping birth triplet lambs at midnight in a rainstorm.

It is often said that the best fertilizer is the farmer's footsteps. This means that being present on the farm and in the fields—looking, watching, and considering—is a crucial component of developing skillful farming methods. Yet this watchfulness demands your time, and it needs to happen at different periods throughout the day. The farm is then a co-creation of the practitioner and nature, existing outside the constraints of a regular schedule. This can be a recipe for endless work.

Because of this, marriages and partnerships can get destroyed on farms. It doesn't matter whether the spouses are both farming or whether one has an off-farm job. Stress abounds in farming: bad weather, employee drama, sick animals, broken machinery, sore backs, and bad weather (did I already mention that?). Just one of these challenges can bring bushels of hardship into the family. Compounding things further, a lot of money and a bunch of assets are usually at stake.

When I reflect back on my twenty-five years in full-time farming, although I have endured a litany of weather-related troubles, as well as physical traumas, the relationship drama between me and my business partners, employees, and family has truly been the hardest part of farming. While relationship issues can dog nearly anyone in any career, I do believe that farming, as an occupation, is particularly stressful on personal relationships. Here is a vital piece of insider advice: Learn early on how to speak your truth without criticism or blame, and learn how to listen without taking offense.

This requires some hard work, and I recommend that you cultivate good communication skills in whatever ways suit you—reading, workshops, training, therapy, or other personal-development work. I started my own individual work with a psychotherapist at age forty-five, and, in retrospect, that was about twenty years late. I cannot overstate the value of having a relationship coach. Your coach will lend you a trained and sympathetic ear, and offer lots of perspective. With no dog in the fight, they can call you out, when appropriate, and also give you practical ideas for how to better listen and communicate. Perhaps most important, a coach can remind you not to take it personally!

Here are more ways to strengthen your personal relationships:

- Become a good farmer. This means learn your craft well, so that the growing part, as well as the business part, is less stressful. This will allow you to use your time more efficiently, leaving more for family.
- Make some sacred time for your love life. For instance, plan a consistent date night or Sunday bike rides. Come up with your own rituals, and stick to them.
- Figure out a way to leave your stresses outside the front door when you come home. Sit in the mudroom for an extra five minutes, do some deep breathing, or turn on some music as you prepare for dinner. A glass of red wine or a cold beer never hurts, either!

- Appreciate your spouse or partner for putting up with you, and with the odd hours and demands of your chosen profession. Express your sincere gratitude on a regular basis.
- Establish hard deadlines every day, so that you can go home and have dinner or play with the kids. There's nothing like having to pick up little Johnny at the babysitter promptly at 5:30 PM to encourage you to get your work done efficiently. The people who work for you will appreciate these regular start and end times, as well.

Special Note About Children

Many of us share a vision of our children growing up on the farm, frolicking amid the plants and animals, enjoying the sunshine and sweet air. This is a worthy dream! Although there is no right way to raise kids on a farm, there is one thing for sure: You can't get much work done while watching children. As I have heard farmer and author Jean-Martin Fortier say, "If you think you can farm and parent small children at the exact same time, you are doing neither activity well."

Successful examples abound where children who were raised on farms have grown up to be wonderful self-realized adults, even if they don't grow up to be farmers themselves. Ask any growers you know for advice about how they managed to juggle farming and parenting. Many families choose to hire more employees so that one parent can manage the household. Other families bring childcare help onto the farm. Some take their kids elsewhere for care—which, for me, was the only option. As a farmer with a husband working off-farm, I had to rely on paid local childcare so that I could keep up my work responsibilities. This wasn't the family-farm fantasy I'd once envisioned, but it worked, and my kid turned out great.

There are also examples of kids being pressed into service on the farm, with poor results. Typically, this creates lifelong resentments and an aversion to farming. The key is to teach the lessons that the farm has to offer your kids, without turning them into worker units. Clearly, this is a balancing act, which evolves as the child matures. It's wonderful to rely on farm kids for some services, but they also need plenty of time to just be kids.

My own son, now twenty-one years old, only began to see the benefits of being a farm kid once he got to college. Compared to his friends, he found that he was the only one who could reliably get things done. He had learned from me and from the work, how to approach a job, how to divide up the tasks and steps, and how to keep at it until it was finished. These skills will serve him well throughout his life. He will most likely not become a farmer, but I've always been perfectly happy with this possibility.

So remember, having kids grow up and help out on the farm is about teaching life skills and lessons, not forcing a career of farming upon them. The latter attitude ultimately pushes more kids away from the farm, usually quite the opposite of your intentions.

YOU AND YOUR COMMUNITY

Funny, isn't it, the idea of having a relationship with yourself? Being self-employed means just that—only you are fully responsible for generating your income. Ironically, one person must in essence become two, simultaneously a boss and an employee. Being an independent farmer is a unique mixture of liberty and responsibility, a duality of checks and balances that forces you to walk a philosophical tightrope. It can feel simultaneously exhilarating and quite lonely, balancing up there all by yourself.

Still, it doesn't mean that you have to work alone. In my case, I spent a lot of time working solo, but that's because I was the main tractor operator on the farm, and most tractor work is done alone. But I had farming partners and lots of employees, so other work got done on the farm without me having to do it or oversee it. This pleased me to no end! I loved having three or more different projects or jobs receiving attention at the same exact time. It was a feeling of victory, virtuousness, and productivity.

I spent most of my career being the crew leader—attending every morning and after-lunch meeting, making on-the-fly decisions in the field as the day unfolded, and scheduling who worked where, with whom, for how long, and doing what. All of this was on top of getting my personal list of jobs done.

Chris Blanchard, an experienced grower and farm consultant based in Decorah, Iowa, rightly teaches in his labor management seminars that managing people is a job unto itself. As a rule of thumb, if you have more than six folks working for you, then keeping them happy and productive is a full-time management job. Many smaller growers don't account for this in their dreams of scaling up. Either you will need someone to help cover your jobs for you, or you will need someone to manage your workers. See how this gets complicated?

Beyond having to ensure the regular farm work gets done, you will be reliant on the generous support of others in your family and community to succeed. You'll need allies. Lots of allies. Again, you will need skilled service providers to fix things quickly and well. You will need ten people to show up for three hours to put plastic on your new greenhouse. Borrowing someone's root digger to harvest garlic once a year will save you hours and hours of digging by hand. One day, before the next supply truck arrives, you will run out of greenhouse soil and will be desperate to borrow three bags to get you through. Every Friday afternoon, you'll need someone to help bunch flowers for Saturday's market. Sometimes, you'll need someone to watch your kids on short notice. And not to be overlooked, you'll need someone to go out with you on a Thursday night for a beer and a movie.

You will be deeply reliant on having workable relationships throughout your farming career. The better those relationships, the happier and freer you will be. So unless your agricultural vision is really just a version of frontier homesteading, or bare-minimum one-person production farming, you must become good at working with others. Clear, consistent, and courteous communication will be primary building blocks of your success.

OUT ON YOUR OWN

I have known a number of young growers to strike out on their own after a few years working and training at others farms. While they wanted to stay close to the community where they trained, many of them had to go two to three hours away to be able to afford land. So, not only were they starting brand-new farming operations but also they were doing it in unfamiliar geographies, without any social network to support them. This is a recipe for increased stress, at best, and outright failure, at worst. If you must try your luck in unfamiliar area, look for a place with cultural indications that you will find folks to connect with—such as a co-op grocery store, a bookstore, a locally owned café, a Montessori school, or a microbrewery.

1. Who else will work on your farm? What relationship models will you employ? If you are already an experienced grower, will you offer a structured learning environment to your staff? What might that look like?

2. What relationships do you currently have with service providers? Are they well tended? What new ones do you need to forge?

3. What other support systems do you have in place for ensuring happiness and success in your personal relationships?

CHAPTER THIRTEEN

THE BEAUTIFUL PARADOX
OF FAILURE

Forrest

The ancient Greek philosopher Heraclitus once wrote, "No man ever steps in the same river twice." Of course, he didn't mean that you can't roll up your pants and step into the Mississippi one day, but not the next. Rather, he was referring to the passage of time, and the ever-changing nature of . . . well . . . nature. Time, like other elemental forces—water, wind, sun—is ever in flux. No matter how much we wish it otherwise, certain things are simply beyond our human control.

Yet, throughout history, people have spent entire lifetimes ignoring this, attempting to distance themselves from nature. You'll find them today, striding waist-deep into metaphorical rivers, boldly shouting "Stop!" as though they expect the current to part midstream. Not surprisingly, their results are usually disappointing. When you know where to look for it, it's easy to spot this sort of misguided intransigence against nature everywhere in daily life.

Agriculture is no exception. Over the decades, I've witnessed scores of farmers try to forcefully re-create the same day, the same season, the same product, over and over. Admittedly, to a certain degree, this desire is understandable. There's comfort in repetition, the reassurance that something will be exactly the same every time you experience it. This is why apple pie, baseball, and *Star Wars* movies are so popular in our culture: They follow the same recipe, the same rules, the same intergalactic explosions time after time. It's probably inevitable that Congress will enact a Star Wars–Baseball–Apple Pie Day, giving us another three-day weekend in September.

This perception of control over nature, however, is one of the core differences between sustainable and conventional farming. Conventional farming strives to create guaranteed results by exerting control over the soil, the ecosystem, the genetic coding of the very seeds themselves. Again, at first glance, this might seem compelling, or even desirable. For example, what's wrong with growing perfect, identical corn from sea to shining sea? Playing this thought process out, however, it's easy to see how it translates into a fixed landscape, an inflexible economy, a reliance on middlemen providing chemicals and machinery. Eventually, it even creates a rote way of life. In mainstream farming, huge amounts of effort and energy attempt to force Heraclitus's river to come to a halt—frozen in time, one perfect production year stretching into eternity.

But with hundreds of millions of acres currently under production in these systems, it certainly seems that human control *can* be forced upon nature. Could it be that Heraclitus was wrong after all?

Fat chance. I've never woken in the morning and stepped into the same farm twice—not even during my days of raising monoculture corn and soybeans. Nature always finds a way to intervene, in spite of an endless array of man-made solutions. For example, with the widespread adoption of crop and livestock monocultures, superweeds and antibiotic-resistant bacteria have recently flourished. Regardless of how one chooses to farm, pests, predators, blights, droughts, blizzards, tornadoes, and hurricanes will always be proof of how little control we actually have. Throw in a flood, and we suddenly discover Heraclitus's river flowing through the middle of our very fields! Far better to acknowledge and adjust to these risks well in advance than to evacuate your sheep at midnight in emergency canoes.

Sustainable farming is built around the expectation that things change, that adaptability and innovation remain paramount, and that failure, when it occurs, is a critical teaching tool. This doesn't imply that you should relinquish your role as a conscientious manager; rather, it means that you should recognize the folly of thinking that you can force full control over nature. With this mindset, you soon learn that challenges and occasional failures aren't things to be afraid of. They are actually welcome components of a resilient operation, and opportunities for learning and growth.

Failures guide you to solutions. There's nothing like finding your flock of goats (all seventy-five of them) loafing at the top of your haystack to remind you to always put the bale elevator away. When you arrive at market and discover that half of your tomatoes are ruined because the crates were improperly stacked, a protective five-dollar sheet of plywood suddenly looks like a godsend. And everyone reaches a point in farming where no sight is more beautiful than four truck tires fully inflated with air. It might only be 10°F outside, and the heater isn't working—but hey, at least you're rolling.

Physical problems like these are fairly easy to conceptualize. What's far harder, however, is dealing with *human* nature—the enigmatic, energetic ebb and flow of personal shortcomings, foibles, and miscalculations. Regardless of good intentions, on a farm, you will inevitably go down the wrong path from time to time. When you do—and especially when your ego gets in the way— you might prefer to face a hurricane than to concede the honest truth.

Sometimes, it's easier to believe that the wind blew the barn door off the hinges than to admit that you forgot to properly bar the door to begin with. Last night's wind must have rattled it loose, or maybe it was the cattle scratching against the nail-studded boards. Come to think of it, the neighbor's kids were jumping in the hay last week . . . maybe it was them!

This is the nature of Heraclitus's river, too, the constant variability of our own lives. Sure, you've barred that door a thousand times before. But this time, when you stopped to admire a red-tailed hawk or answer a text message, the current shifted a little. You got distracted, and you forgot. Just as no two days are ever identical, you are never the same person from one day to the next (and who would want to be? Boring!). The sooner you accept that mistakes happen—and that this is all part of the normal, healthy flow of daily farm life—the sooner you will become more resilient and productive.

This is the beautiful paradox of failure. It will arrive in all shapes and forms, despite your most strident efforts to prevent it. Yet without it occurring, you rarely improve or move forward. Just like a toddler making his first wobbly steps, in farming, you are guaranteed to fall backward, sideways, or occasionally straight onto your face. This, however, is how you learn, how you build your strength and your balance. It's these first steps—and subsequent hard landings—that will one day allow you to run with confidence.

The river of change will always flow. Roll up your pants, and go wading.

MY OLD PROFESSOR, HENRIETTA CHICKENHEIMER

In one way or another, I can attribute practically every lesson I've learned about failure to chickens. It's yet another paradox, isn't it? Impetuous, scatter-brained, and, well, plain chicken, poultry has been one of my greatest teachers. I've spent four decades raising practically every variety and type of chicken, duck, and turkey, and any accomplishments I've accrued have been, without exception, prefaced by failure.

Poultry dreams always start so auspiciously. Imagine a thousand ruby-red hens photogenically scattered across a meadow, scratching for bugs and nib-bling clover, the midday sunshine beaming, a gentle breeze bending the blue-grass. The hens sip spring water, and rolling feeders provide locally grown corn and wheat. And in each nesting box, a dozen or more perfectly formed, freshly laid brown eggs are ready to be gathered.

So what's the problem? As much as I love them, chickens are essentially defenseless, feathered nincompoops. If you think that there's anything natural about one thousand chickens running around in a field, ask yourself this: When was the last time you saw a single wild chicken, much less one thousand?

Nature reminds me of their fragility every day by constantly testing my systems. Despite installing many protections, I've had chickens preyed upon by coyotes, foxes, mink, weasels, skunks, raccoons, opossums, dogs, snakes, cats, kestrels, hawks, owls, and falcons. I once watched a high-strung hen fly over my protective netting, sprinting toward a destination unknown, only to be nabbed mid-stride by the most patriotic predator of all, a bald eagle. In one pendulum swoop, she was lifted skyward, doomed. Three melancholy feathers, sifting to the ground, was all that was left. Give me one thousand chickens, and I will recount fifty ways that they unexpectedly expired over the course of three hundred and sixty-five days.

Fifty is no arbitrary number—it works out to 5 percent. If it's not preda-tors or some brutal weather event, then the chickens somehow manage their demise themselves. After a lifetime of raising poultry, 5 percent attrition is the number that nature reliably manifests. As such, this is what I plan for. Of course, I take every precaution to ensure that my hens are content, safe, and free-ranging outdoors despite the season, deep snow notwithstanding.

The lesson, though, is clear. When nature intervenes, I'm the one who must figure it all out. This is my duty, as well as my mission. When failure inevitably occurs, instead of giving up, I view it as an opportunity to make corrections, to improve, and to test the results. This, again, is the beautiful paradox of failure. In order to make progress, weakness must be revealed. When it is—and when you respond accordingly to correct it—then next year's growing season invariably improves. My hens are happier. I'm happier. My profits increase. And my customers really, really enjoy those free-range eggs.

The Cost of Avoiding Failure

Different agricultural methods reflect different approaches to failure. For example, outdoor, free-range poultry systems operate in stark contrast to 99.9 percent of how most chickens are raised. Mainstream agriculture seems to reverse engineer the process, focusing on all that can go wrong outdoors, then attempting to correct these problems with manufactured solutions. This effectively eliminates nearly all natural conditions. Raised indoors their entire lives, laying hens are confined to vertically stacked cages, with sloped floors that allow eggs to roll onto conveyor belts. Temperatures and lights are mechanically regulated. Confinement buildings eliminate the threat of predators and inclement weather. This is a classic attempt to make Heraclitus's river come to a complete stop.

While this is certainly a reliable way to produce eggs, these methods of preempting failure come with their own accompanying consequences: animal-welfare concerns, manure-management requirements, mediocre product quality, air pollution for neighboring communities, and low commodity wages for the producer. The goal in farming, we believe, should always be to achieve profitability as well as a superior product, alongside joy and satisfaction for all involved. From our experience, working *with* nature—so long as we learn and improve from our failures—offers a clearer path toward these goals than working in *opposition* to nature.

SEASONALITY IS FOR THE BIRDS

Many years ago, I was at a farmers' market on a drizzly early-April morning, lamenting to a fellow producer about my most recent poultry problems. That year, instead of starting my chicks in May as I usually did, I had tried to get a jump on the season by warming up my brooder in March, and turning the birds onto pasture with the first flush of green grass. But that year, a solid week of frigid rain and sleet had been nearly disastrous to my half-feathered chicks. I had spent the past seven days rigging up mobile heat lamps for six different shelters, barely able to save them.

Listening patiently, my vegetable-farming friend replied, "We used to try to get a jump on the season as well, but we lost too many seedlings due to the cold. Now, instead of trying to rush the season, we just wait till it's warmer." He shrugged. "Turns out, the customers will buy it when we have it, and it costs us half as much to produce."

This was a revelation to me: Work within the seasons, grow more for less money, and keep your customers happy along the way. My friend's commonsense advice forever changed how I viewed production.

ENTERPRISE FAILURES: FIREWOOD AND DUCKS

When I was just starting out, I had far more time at my disposal than I did money (recall our discussion of time versus money in chapter 2). Because of this, I tried everything imaginable to generate revenue, by leveraging my hours instead of my extremely limited nickels. Early on, I cut firewood for an entire year, delivering cord after cord of gorgeous red oak, hickory, and black walnut. This served the dual purpose of generating revenue, while also cleaning up my pastures, and only required borrowing my dad's chainsaw and burning through five dollars worth of gas daily. It seemed like a great plan.

The failure was that I didn't account for the heavy toll the firewood deliveries would exact on my little Toyota pickup truck. At the end of the year, my transmission was shot, and the profit I had made was subsumed by a $3,000 repair bill. The following spring, it was back to the drawing board. I quickly learned, however, that hauling firewood wouldn't be part of my future plans.

Raising ducks was much the same experience—mostly requiring my time, with negligible material cost. For the price of some nails, old boards, and a handful of heat lamps, I was able to construct a brooder for my ducklings. Thrifty and self-reliant, I felt like Henry David Thoreau surrounded by a hundred garrulous, flat-footed admirers.

The problem arose, however, when I moved them out onto pasture. Unlike chickens or turkeys, which prefer to remain dry, ducks are waterfowl and relish being wet. I soon discovered that they splashed all of their water onto their backs, playing instead of drinking. This was fun for them—that is, until the water ran out, and they were soon thirsty. I thought I had solved this problem by installing automatic watering systems, but this only exacerbated the problem. With unlimited water, the ducks soon spent the entire day splashing, eventually creating huge puddles. Imagine my consternation when I walked across the pasture to find my field a swampy mess, with one hundred ducks delightedly quacking as though it was the most glorious day of their lives!

Turns out, what was good for the ducks was terrible for my pasture—we referred to the muddy, compacted ponds in our field as quackmires. It actually took several years for the field to recover, costing me far more in future production than the modest profit I made from selling the birds. The lesson? In their own way, the ducks tried to tell me that they didn't belong on my farm. Yes, I could technically raise them successfully, but pairing their nature to my system had been a failure, and I hadn't matched the land to its suited use. It is far better to focus on enterprises that provide natural synergy than to force a fit.

EXPANSION FAILURE: THE FABULOUS FOOD TRUCK

One of my biggest failures occurred, ironically, because it was too successful. Remember back in chapter 8, when I emphasized adding value to what you grow the most of? In 2011, I had the ambitious idea of starting a food truck in Washington, DC, turning my ground beef and sausages into cheeseburgers and breakfast wraps. To do so, I purchased a used sixteen-foot fish-and-chips restaurant trailer from a Craigslist ad in Baltimore, and spent an atrociously filthy week with grease up to my elbows, scrubbing it from top to bottom. Finally, as a finishing touch, I branded it with my farm's logo.

Its popularity was immediate. Only operating Saturdays and Sundays during our four-hour farmers' markets, the truck grossed around $3,000 per weekend. After approximately $1,000 in expenses, that left a cool $2,000 in profit, come Monday morning. My future appeared to be paved with grass-fed cheeseburgers and sustainably crafted sausage wraps. All that was left to do was carry bulging bags of cash to the bank each week—and perhaps enjoy a cheeseburger myself from time to time.

So what was the problem? Too much success, far too fast, was our undoing. I spent that first year so focused on building the food-truck business in the city that details back at the farm began to suffer. Remember the three golden rules of farming (chapter 11, page 175)? I found myself stretched too thin to follow them. Before long, broken fences went unfixed, leaky hydrants were left leaking, and tools became scattered. In short, I was so focused on making hamburgers that I neglected producing the hamburger-giving beef!

Surely there was a solution. Asking for advice from mainstream business-people—experienced individuals whom I respected, but who weren't farmers—I was informed that I only needed to hire a manger, and my problems would be solved. In hindsight, this was some of the worst advice I've ever received. Within a year, I went through three managers. The problem? Though perfectly nice people, and very well paid, these managers were far less invested in the business than I was. I soon found myself following along behind them, tidying up oversights, filling labor gaps, making emergency trips for extra mustard or paper towels. Success was running me ragged!

The final straw was when my third manager arranged late-night food service with Georgetown University. On the very first night, he got the truck stuck in the tunnel in the middle of campus, and called me in a panic. The sign read eleven foot six; the truck was eleven foot eight. I drove an hour into Washington, backed the truck out of the tunnel, and politely—but firmly—shut down the business that very night.

After that, I returned my focus to what mattered most: doing what I loved, and doing it *well*. Popular and profitable as it was, the food truck offered neither of these important things. It took a "successful" failure to help me clearly see what was important in my life, as well as to remind me what I truly loved to do.

SYSTEMS FAILURE: THE MYSTERY OF THE SHRINKING LAMB CHOPS

As Ellen explained in the chapter 12, division of labor on a farm can be a beautiful thing—so long as everything goes smoothly. Like most farmers, I don't have a USDA butcher shop on my farm, so, for many years, I used an abattoir in Maryland for all of my processing needs. But they were constantly busy, so when a new processor opened up just a little farther away in Pennsylvania, I decided to divert my lambs to the new location. The new butcher was incredibly helpful and friendly, bending over backward to accommodate my production schedule. The first year, everything seemed to be going great.

Gradually, however, I began to notice something curious. For years at the old butcher shop, my lambs had always yielded 40 percent of body weight. This meant an average one-hundred-pound lamb provided a reliable forty pounds of product, including packs of lamb chops, each weighing approximately a pound. Yet with this new shop, the chops always came back closer to three-quarters of a pound—and the other cuts generally weighed less, as well. I was only getting about thirty pounds of total cuts per lamb.

Some quick math showed that it was the same number of chops, so I had no concerns that any product was missing. Yet at $20 per pound, over the course of the season, a quarter-pound loss per pack translated to a whopping $6,000, just in lamb chops alone. This didn't even account for the other cuts.

This mystery was solved after I hired a new staffer, someone who had previously worked in a butcher shop. We were packing for the farmers' market when she picked up a pack of lamb chops and studied them intently.

"Why do you dry age your lamb for so long?" she inquired.

Dry-aging was a long-established part of our system, a process where meat is hung in a cold room for ten days, to tenderize before processing. "What makes you ask that?" I replied.

"Because," she explained, pointing at the package, "these chops have been aged way too much, and they've lost all their moisture. Look. They're shrunken."

I hadn't noticed this visual discrepancy before, but she was right. And that's when it finally hit me. The new butcher always took a month or more to get my product back to me, as opposed to the two-week turnaround from my other processor. I had always assumed that the new butcher had simply forgotten to call me when the product was ready. Now I realized that the meat was hanging this entire time, desiccating to something just north of jerky.

Failure, it turns out, arrives despite your best intentions. And what a failure this was! Extrapolating the math, I soon realized that $6,000 in lamb chops was a drop in the bucket. When I accounted for all the cuts, a 25 percent loss on one hundred lambs translated into nearly $20,000 over the course of the season. My profits for that year's lambs had quite literally evaporated. And that didn't even account for the fact that I had been unwittingly selling products that weren't up to our standards—who knows how many customers I lost as a result?

EQUIPMENT FAILURES: PAYING FOR YOUR MISTAKES

I can't specifically count how many times I've smashed my thumb with a hammer, ripped my britches on barbed wire, or pratfallen while trimming sheep hooves. I *can* tell you, however, about the one time I engaged the haybine in transport mode, ripping apart the drive shaft to the tune of $2,500. Pulling an overloaded wagon of round bales, I remember gunning my engine to the point where motor oil suddenly exploded out of the crankcase, a $4,000 miscalculation. My truck was down for three weeks.

When I bought my first walk-in freezer from an out-of-business catering company, they insisted that it was in perfect working order. After hauling it back to the farm and reassembling it, I discovered that, indeed, it was in perfect working order . . . as a *refrigerator*, not a freezer. This "freezer" had cost me $3,000. But to make it a real freezer, it required an additional $3,500 worth of mechanical retrofits.

Each of these lessons was due to the fact that I didn't fully understand the limitations, capabilities, or engineering of some very expensive equipment. By not taking the time to properly educate myself, each failure ended in thousands of dollars of costly repairs or modifications. Far better, in my opinion, to rent equipment and gain experience, or to ask the advice of a professional before making major purchases. Better yet, visit other farms or job sites to see machines in action, before you decide to jump in feetfirst.

Owning useful stuff is fun, but breaking it stinks. Paying for it—especially when you know you could have avoided it—is one of the worst feelings of all.

SMALL FAILURES, BIG-PICTURE LESSONS

Paradoxically, failures such as these have usually taught me far more than my actual farming successes. Each experience was in turn humbling, discouraging, and even maddening. But this is what you sign up for as a farmer; it's the natural flow of a living, biological business. For every so-called perfect day on a sustainable farm, rest assured that a so-called failure day is waiting just around the bend. It's what you *do* with failure when it arrives—how you react, how you respond—that ultimately shapes you into the farmer you wish to be. As the author F. Scott Fitzgerald famously remarked, "Nothing any good isn't hard."

Still, there's hard, and then there's really hard. I've saved the following list for last, because these were some of my greatest, most impactful lessons, the failures that pushed me to the very limit.

Staffing Failures

Placing full reliance on outside help to run your operation—while often necessary to maintain a certain scale or a certain type of production—comes with inherent risks. Over the course of several decades, I've had people quit, move on, or take other jobs for practically every reason in the book. Sometimes, frankly, they seem to disappear into thin air. To be sure, I've mostly had wonderful, conscientious employees. Sometimes, life just gets in the way, and people need to change jobs. On the other end of the spectrum, however, I've had staffers get arrested, or I've had to fire them after they lied to me or stole from me. There was even one who went completely unhinged and threatened to murder me. I barely slept for a week!

I've been at this long enough to know that, eventually, regardless of the reason, the morning will eventually arrive when someone doesn't show up, and this will have a tremendous impact on my ability to run my farm. It might be next year, or next week, but it will happen. Yes, I have a great staff, and we work extremely well together. But again, this sort of failure can occur for numerous, completely legitimate and understandable reasons. Staffers get sick. Weather emergencies can prevent travel to the farm. A traffic accident might occur. There could be a death in the family. Still, the farm must go on. After decades of these experiences, my ultimate takeaway is this: I can't reasonably expect anyone but myself to make my farm run. If things go well (as they usually do), then great. But at the end of the day, I know that I'm the one ultimately responsible for ensuring that the show goes on.

I can't ever allow a staffing failure to shut me down. As such, all of my systems are intentionally geared for one person to operate, solo. Usually, that's one of my two full-time farmhands (besides me, there are two staffers on the farm), and on any given day, it all works out great. Cattle and sheep are rotated to fresh pasture, the eggs are gathered three times daily, and the pig waterers and feeders are inspected. In short, management of hundreds of cattle, sheep, hogs, and a thousand laying hens can all be orchestrated by a single person.

Don't get me wrong. A lot more gets accomplished when all three of us are working together, and this is how my farm usually operates. But having a system that I can run by myself, if need be—at least for a day or two until the staffer returns, or until I can hire someone new—is invaluable.

By intentionally building simplicity into your systems, it provides additional flexibility, too. When you get the flu, break a leg, or have a family emergency, it's critical to know that the farm can go on for a day, a week, or even a month, humming along with a lone, experienced staffer. Again, this certainly isn't ideal, and not much other than basic chores might get accomplished. But the fact that it's possible is what matters.

Of course, definitions of "simple systems" will certainly vary from operation to operation, and what might work for a livestock farm, for instance, will never directly translate to a high-volume vegetable farm, or vice versa. But creating a system that can maintain production, while remaining resilient to the X factor of a staffing breakdown, is a fundamental step toward sustainable success, regardless of enterprise.

Life-Threatening Failures

Nothing taught me faster to be a better tractor operator than popping a wheelie going up a steep hill, my front tires soaring eight feet off the ground. Fortunately, the wide tiller on the back bounced me earthward and kept me from flipping entirely. I've had a one-ton post-driver crack its welds, delivering a killing blow to where I was standing just seconds before. Another time, a seven-hundred-pound hog chased me straight up an apple tree, fully intent on eating me for dinner. Caught in sudden storms, I've witnessed mighty sycamores split in two by lightning, while inexplicably, I was lucky enough to avoid electrocution.

Farming is more than just hard. It's statistically one of the most dangerous occupations on the entire planet. Accidental death, or a life-changing injury, is a real and ever-present threat. In each of the above instances, I was fortunate enough to avoid the ultimate failure—failure to protect my own life. Each time, I learned to be more careful, to pay closer attention, and to stay light on my feet. Far better for failure to be a learning opportunity than a final blow.

This doesn't mean that we need to approach farming with fear or overdue concern, but rather with proper preparedness and diligence. Each day, I wear appropriate gear on the farm, which I affectionately call *my armor*: full-length, stiff canvas pants, steel-toed boots, and a sturdy belt with a multi-tool in a holster. Brambles scratch. Cattle stomp. Wasps sting. Snakes bite. These are minimum daily considerations. You have to take sensible steps to protect yourself from daily injury.

Big-picture preparedness is crucial. Inspect tractors and trucks regularly, make sure that livestock systems are in safe working order, and follow the proper precautions when using power tools, machinery, and other dangerous equipment (yes, this means *always* wearing goggles and earplugs). I nearly lost my leg one afternoon, when a volunteer haphazardly swung a large branch into my running chainsaw. The blade tore a ten-inch gash in my jeans, but miraculously, I jerked it away a millimeter before it sliced open my thigh.

We'd all like to believe we're invincible superfarmers garbed in plaid and denim in lieu of tights and a cape. The fact is, of course, we are all thoroughly human. I prefer to remain so for as long as, well, humanly possible.

Failures of Faith

My most humbling and deeply despondent periods as a farmer have always been during prolonged summer droughts. Hundreds of times from my front porch, I've watched a thunderstorm slide just to the north or south of the farm, the downpour a scant mile away, but nary a drop reaching my simmering, parched fields. When a drought really sets in, I sometimes even forget what rain feels like. In my bleakest moments, I start to think that maybe, this time, it's forever, that the rains will never come again, that the farm is going to fail. It's a hopeless, helpless feeling.

The rains won't come today, and maybe not tomorrow. Maybe not for an entire month. By now, however, I know that life-giving rains will eventually arrive. They always do. Just as waist-deep snows always melt, or a howling wind finally blows itself out, nourishing rains will at long last slake my thirsty fields. I just can't predict when.

Because of this, it's what I do in the meantime that's of the utmost importance. To combat droughts, I leave ample biomass to cover and cool the soil, fostering deep roots for access to subsurface water. Years of rotational grazing have increased organic matter in the fields, microscopic carbon sponges that—when it does finally rain—sequester thousands more gallons of water than the ground held in the past. This extra available moisture allows my crops to ride out days or weeks of drought that they previously couldn't have endured. Years of advance planning and subsequent execution has had an inestimable impact on my profitability and my productivity, as well as my personal happiness.

This is all a way of explaining that the greatest test that any farmer experiences is a failure of faith. It cuts us to the core. With experience, we learn to have faith in nature, faith in our systems, and faith that something far greater than ourselves guides us. Without deep commitment to each of these, when things get really hard, it's much too easy to throw in the towel.

It's only when you're pushed to the very brink of failure that you are truly tested, that your mettle as a farmer is fully revealed. Extreme challenges yield extreme insights. Painful in the moment, yes, but, ultimately, positive. It's a sublime paradox.

Heraclitus's river is always flowing, always challenging you. What at first seems like a lack of control is actually your active, intentional participation in life-bringing change. When you work with what nature provides—and accept that failure will test your systems—you steadily strengthen your farm. The sun will always shine again, and the rains will always come. Understanding this, you gradually dispel the illusion of human control, appreciating the river of time for all its splendor.

CHAPTER REVIEW QUESTIONS

1. Think back to one of your biggest failures in life. How did it shape you into the person you are now? What did you learn? Does it still feel like a failure, now that time has passed?

2. How important is it for you to maintain a sense of control? If this is a strong need, can you cope with the highs and lows that come from working with nature? Or, conversely, if you're less controlling, will you be able to instill more organization into your life?

3. Have you ever been in a situation where you were physically and emotionally challenged, and where self-reliance was required for a successful outcome? What insights about yourself (and the larger problem) did this reveal to you?

CHAPTER FOURTEEN

CASHING YOUR
LIFESTYLE PAYCHECK

Ellen

We've talked openly and repeatedly about the concept of return on investment. What do you get back from having invested in farming assets—the land, the buildings, and the equipment? Certainly, you must leverage those assets to create positive cash flow, as well as profit, in order to stay in business. But growing food is a tough path to what most people consider serious riches. Instead, we aim for modest comfort in our standard of living, topped with tremendous satisfaction at jobs well done. Rather than a two-hour-a-day commute, we wake up right where the work is—boots on and out the door, we start the day. Beyond money alone, what we get in return for our investment is what we call the lifestyle paycheck. This paycheck comes in many shapes and sizes. While it can't be deposited at your local bank, it pays you over and over, in myriad ways.

WORKAHOLIC, OR FULLY INTEGRATED?

For many of us, the allure of farming as a career is represented by the dreamy vision of a dew-soaked field as the sun rises, or witnessing the first miraculous steps of a newborn lamb. Indeed, these are the moments that we yearn for as beginners and then cherish as practiced growers. But realistically, many of these moments do not happen between 9:00 AM and 5:00 PM, Monday through Friday. Rather, they are likely to happen at dawn or dusk on any day of the week, whether it's Memorial Day or not. Going against the grain of our mainstream culture's insistence on keeping work at work, and your personal life at home, farming mixes everything all up into one complicated, integrated ball of work-life balance.

I know of only a handful of farm owners who manage to keep "normal" business hours. These rare producers have been in business a long time and have very capable, permanent staff to keep the farm humming along, almost on autopilot. For the rest of us, bookkeeping and other office chores often happen when it's dark outside, the livestock get checked after dinner, and farmers' markets are worked on weekends. A farmer's life and work are completely intertwined.

While this might seem intimidating at first, this can actually be one of the biggest rewards of choosing the farming life. The benefit of working early and late in the day might be your ability to attend your child's school play at 1:30 in the afternoon, or taking a power nap at 3:45, or attending a 4:00 yoga class in town. This is the freedom and flexibility that comes from being the one to set the schedule for the day, the week, and the season. While the weather sometimes toys with our best laid plans, we alone get to decide if attending a Sunday farmers' market is worth missing Suzie's soccer games. As stated before, choice is good. Within the bounds of your geography and market, the farm is yours to create. You aren't "working for the man"; you are working for yourself.

FAMILY MATTERS

We've considered the topic of family in chapter 12, primarily in terms of how to keep the farm from negatively impacting those relationships. But what are the rewards for the farm family? Farms teach life lessons to all involved—about the most basic preciousness of life and death, birth and renewal. To continually and viscerally bear witness to the passing of the seasons is one gift the farm gives to its participants.

Another invaluable lesson that farm work teaches us, especially our kids, is that it doesn't really matter whether you have a headache, feel lazy, or are sad—some things just *have* to get done, no matter what. The biological needs of plants and animals trump your mood. The greenhouse must be watered, the cows milked, and the pigs fed. These life necessities are absolute and help us remember the true priority of biology over attitude. In my experience, allowing the work to call out my best effort ultimately yields fewer headaches, more energy, and better moods. The same goes double for kids. The farm itself instills a biological balance in the family.

When your child is asked at school, "What does your mom do for a living?" he will have a litany of real-life answers. From a nuts-and-bolts, daily-chore standpoint, farm work is not mysterious or baffling. It's mostly comprised of understandable, explainable tasks, with obvious and tangible results. Your child can proudly represent what you do all day. He can even hold up an apple, carrot, or a ham sandwich from his lunch box, and show his classmates what you do!

Being able to actively observe you working is a gift to you both, garnering your child's respect and admiration. And, likewise, as you teach and coach your child to take part in small jobs on the farm, you, too, can admire and respect his contribution and effort. This currency of mutual esteem fosters strong, healthy family bonds. I have found this payback to be of tremendous, lasting importance in my relationship with my son. While a farm is surely not the sole setting for this kind of bonding, it is a uniquely ripe one.

HEALTH AND WELLNESS

At its simplest, farming is a lot of good, hard work. It's a genuine calorie-burning activity, suited for the way our bodies are engineered. Yes, you'll get tired, but this type of fatigue is remedied

KIDS INSPIRE NEW IDEAS

When my son turned eleven, I decided he no longer had to go to camp during the summer days. He chose a couple of programs that he was especially interested in, but mostly he was going to stay home. I needed a new tool to help me stay connected to him while I was out in the fields. So I decided to get a golf cart for our family. He absolutely loved it!! He would call me on the phone asking if we needed anything in the field, as he was chomping at the bit to have a reason to take it out for a spin. He would ride back and forth from the house to the field multiple times a day. He could easily join the crew for a few minutes, or even an hour of work, and then ride back home when he got bored or tired. He especially loved to bring us snacks. There are countless, mutually beneficial ways for family members to engage with the farm. We encourage you to get creative!

by rest, hydration, and nutrition. You have to take care of your main production machine—your body, which includes your brain! Plenty of sleep, lots of water, and great meals will go a long way toward maintaining a well-tuned body.

However, with the exception of throwing hay bales every day, or shoveling stalls filled with manure, most farming doesn't offer much in the way of sustained aerobic activity or flexibility training. So unless your enterprise allows you to get your heart rate up on a daily basis, making time for yoga, stretching, and aerobic sports is a must. I know of a number of farms that schedule daily yoga into the calendar. Even a five- to ten-minute calisthenics routine right after the morning meeting helps get bodies ready for a full day of hard work and heavy lifting. As with any other workplace goal, creating a culture around it makes it habitual and attainable. A health-and-wellness attitude will encourage you and your whole team to take good care.

I regret not doing enough to ameliorate the effects of long hours spent working in uncomfortable positions. My once-a-week yoga class was nowhere near enough to counteract the day-after-day strain of bending over to harvest produce. Thus, low back pain became a constant limiting factor for me.

So take it from me: Preventive body care is way more effective, and much cheaper, than remedial measures. Your youthful bounce fades without your even noticing it. Figure out how you and your team can make good self-care part of the daily farm routine. That way, you'll be able to enjoy the beautiful paycheck of farm life, instead of spending that capital on the chiropractor!

Around We Go

It's one thing to acknowledge that all of these wonderful lifestyle benefits come with farming. But it's another trick entirely to actually enjoy and appreciate them. As much as your farm can be an inspiring and motivating place to work and live, it can also be filled with daily, unexpected challenges. Of course, the beauty exists despite all the problems, and the problems become lost among the beauty. But, goodness, it sure can be hard to see it objectively. It's typically doom and gloom on one end, or sunshine and butterflies on the other. The truth, naturally, lies somewhere in the middle. A career in farming is filled with plenty of wild, emotional bounces—daily, hourly, sometimes even minute to minute.

Luckily, you are blessed with the perfect remedies for these mood swings—nature and time. Call upon the grounding resources surrounding you—clean air, vibrant plants, and healthy soil can set a strong foundation for good mental health and happiness. Many people work indoors all day, disconnected from the healing powers of nature, so take full advantage of your outdoor existence by looking up at the sky often, lying flat on the ground sometimes, and noticing the smells and sounds all around you.

A farm connects you to the cyclical nature of time. Your work is deeply intertwined with the repeating cycle of seasons and the circadian rhythm of each day. The dependability of these cycles provides comfort and opportunity to farmers. Each day brings new work—the same chores, yes, but also another chance to do better. I always loved the cleansing aspect of winter, and the chance to restage a new farm dance each spring. The coming of another season, far from feeling repetitive, has always inspired me anew—as a beautiful clean slate upon which to create a successful farm. It's both steadying and invigorating to match your efforts to the progress of time on your farm.

NOBLE CAUSE

Next to being outside almost every day, one of the biggest lifestyle paychecks comes from knowing that growing food is a good and noble pursuit. While so many people in our present culture find their work to be soul-sucking, agriculture provides deep satisfaction in and of itself. Even if the budget is tight and the hours sometimes long, the honor of providing sustenance to our communities energizes and sustains us. The fundamental biological necessity of food infuses the work with meaning. Taking care of plants, animals, ecosystems, and, ultimately, people is important and valuable work.

One of the most striking and original ideas of the CSA movement was that each person has a basic human right to have access to enough land to feed oneself. And, building on this premise, if a person should decide not to farm that land themselves, they must choose who will farm it for them. That was the basis for the first American CSA farms—a contractual agreement of mutual support between an eater and a grower. To me, this highlights the sacred, communal, and elemental nature of farming as a primary and noble undertaking.

ECONOMIC BENEFITS

Numerous cost savings are afforded to home-based business owners. Office space, cell phones, internet access, and the like become deductible business expenses. Once good rain gear, work boots, and sturdy pants have been purchased, a farmer's clothing budget comes close to zero. Expensive haircuts, makeup, and manicures suddenly fall off the monthly-expense radar screen; your livestock and your employees won't give a fig about your having a bad hair day. The farm atmosphere distills life down to what really matters most—health, relationships, connection, and joy.

As for food, well, luckily you are surrounded by it. And not just any food. You grow the freshest and tastiest ingredients on Earth. Within five minutes, you can enjoy a sun-ripened tomato, a crispy carrot, or a juicy peach. Talk about getting paid to do what you love! And as for what you can't grow or choose not to grow, you can trade for it at market or with a neighbor.

This abundance is worth real money. Your food expenses will plummet. Your restaurant budget will likely decline, as well. Once you get used to high-quality, fresh ingredients, you'll find dining out to be markedly less satisfying, if not downright disappointing.

That said, I know growers who have found it difficult to make time to cook as much as they want to. Some farmers, finally so frustrated with not enjoying the literal fruits of their labor, decide to hire chefs to come in twice a week to prepare dinners for them, and five lunches a week for all employees to enjoy. The cost is palatable (pardon the pun), and the payback is overwhelming. Remember, the best sales pitch for your products comes from a satisfied customer—this includes you and your workforce. Take full advantage of the bounty of food around you, and share it as liberally as you can afford to.

BRINGING THE JOY

The counterbalance to the freedom of being a self-directed farmer is that, ultimately, you are responsible for pretty much everything that happens on the farm. Sure, there's always the weather, or Mother Nature, to blame for scads of minor snafus. But for the most part, you are the creator of your own little universe. The farm will have an aura about it—a palpable energy—that you create. It is up to you to manifest joy in as many parts of the farm and in as many minutes of the day as you can. When you pull this off on a regular basis, your lifestyle paycheck is plump with riches.

Try to make having fun a foundational principle on the farm. Few in the farming community share this secret, but here it is: Farm work doesn't have to be boring, bleak, or monotonous. The difference between drudgery and joyful progress is mostly attitude. As any good parent knows, when you make doing the dishes the fun activity in the household, all of a sudden, the kids want to take part in it. People are attracted to positivity and laughter. There's plenty of room for boisterous laughing and doing a jig on a farm—there are no corporate honchos to avoid and no conference calls to interrupt.

Intentionality matters here. At our farm's end-of-season meeting each winter, we started with going around the table to express what was fun for us that year, and what wasn't. If we could make adjustments to our systems or crops to remove the no-fun parts, we would. Nobody liked picking okra, so after determining that it was only marginally profitable, we finally stopped growing it. Sometimes, just listening to someone with a positive view of the offending chore can help adjust an attitude problem. For example, I love to pick green beans (I'm a weirdo, I know), but I've helped many a farmhand to pick faster and be happier by framing the job like this: "Instead of thinking of the row as just an endless, tangled mass of beans and leaves, take it just one plant at a time, and pick the beans on the front side of the plant. Then, bring it toward you, and pick the back half. Then, that one is done."

Other times, a solution may be more involved. For example, I really disliked dragging hoses and dealing with kinks. So the easy solution was to buy really good hoses that keep their shape and last a long time. The more foundational solution was to install a second hydrant in the middle of the greenhouse, which meant shorter, lighter hoses to move around. Making these improvements all over the farm adds up to a big difference in the happiness quotient.

I will admit that I am being loose with my definition of "fun." Perhaps satisfaction and pleasure are other facets of the same feeling. When I brainstorm what's fun for me in farming, here's a stream-of-consciousness list of activities:

- Using the vacuum seeder in the greenhouse
- Riding around in golf carts

- Listening to music while bunching flowers
- High-fiving each other at the end of a row
- Throwing a Wednesday potluck lunch
- Enjoying deep discussions in the bean patch
- Collecting eggs
- Hearing customers rave about our food
- Eating tomatoes right off the plant
- Having the PTO (power take off) shaft click into place on the first try
- Discovering the knives have been sharpened
- Finding the perfect number of CSA bags for each drop site
- Opening a market van full of fragrant produce at 4:30 AM on a Saturday
- Going home from market with only empty containers and a wad of cash
- Staying dry in my rain gear
- Completing hard jobs quickly with lots of help
- Planting on the waterwheel transplanter
- Driving straight rows
- Counting money
- Opening CSA registration to a stream of eager members
- Trading vegetables for meat and fruit
- Savoring cold watermelon in the field on a hot day
- "Popping" the string of clean mulch bales with no thorns inside
- Making delicious meals out of "seconds" produce
- Talking in silly fake accents in the field all day
- Watching my workforce get stronger and faster
- Punching out for Friday "beer o'clock"

A few common themes jump out at me from this list. Food and beverage certainly figure prominently. But more important, we can derive fun, satisfaction, and pleasure from almost any activity where things go well, when equipment works, or when there is positive team spirit. These are all forms of success. And guess what? *Success is fun!*

The Kid in the Candy Store

Sometimes, of course, there can be such a thing as too much fun. Forrest shares a story of this. As a young farmer, he experienced long jags where fun shimmered on every hillside. One hot summer, he imagined a fleet of rolling chicken houses dotting his pastures, where laying hens could free-range on fresh clover, fertilizing the soil as they went. How exciting! It would be a virtual poultry paradise. So he dreamed these structures into existence, frenetically constructing them one hot summer week, and—*poof!*—just like that, his dream came true.

Next, he saw an opportunity to raise even more chickens. He'd process them on the farm in a butchering facility he had built himself. And, so, after a few truckloads of concrete, and a trip up to Pennsylvania to pick up a stainless-steel scalder and plucker—*poof!*—he found himself processing six hundred chickens every other week, all summer long.

Not to lose out on more "fun," he started raising turkeys, building a fleet of pastured gobblermobiles, and soon processed five hundred birds in advance of Thanksgiving. Of course, while he was tending to all of these chickens and turkeys, it occurred to him that, if he just put his mind to it, he could also be raising several hundred goats. So he did! And while he was doing this, he also decided to start raising ducks, because . . . well . . . why not, you know? Then he saw an ad for something called Barbados blackbelly hair sheep. Of course, that sounded fascinating, so he drove his trailer an hour south and picked up several dozen wild-eyed creatures that looked more like diminutive antelope than actual sheep. Fun!

It was about this time that someone gave him a donkey; a few weeks later, a neighbor dropped off two horses for him to watch for a month while they went on vacation. This, of course, was on top of maintaining two hundred cattle, several hundred hogs, and all the while attending six farmers' markets every weekend in order to actually make a paycheck. This was all still fun, right?

Poof! His energy was gone. Forrest was the proverbial kid in the candy store, wide-eyed at the array of sweet treats, gloriously displayed in all directions. Be a sensible person and know that there's only so much you can bite off before you're unable to chew.

It's true that his passion was driving him to worthy pursuits, and potentially fun ones at that. Fun, that is, until his goats broke through his fence each morning for a month, and wandered several miles away. Fun, until he learned that plucking a duck takes twenty times as long as plucking a chicken. Fun, until the horses mercilessly stampeded his cattle, the blackbelly sheep used him as a springboard while he trimmed hooves, and the donkey administered a bruising bite to his rump the moment he turned his back. Turns out it's hard enough to perform one enterprise well, much less twelve.

By all means, find your fun wherever you can. But remember, "the opportunity of a lifetime comes once a week." In other words, the world is filled with endless possibilities, and with experience, you'll grow to realize that there's rarely any special urgency to act right away. Balance your impulse to seize the day with some thoughtful research. Slow down enough to take a good look around. Sometimes, you already have all the opportunities you need right there in front of you. Cashing your lifestyle paycheck means being able to *enjoy* the fun when it occurs. Remember that too much fun, ironically, can mean no fun at all.

Artistic License

It's difficult to overstate the immense satisfaction and pride that comes from nurturing something from nothing—whether it's planting a seed the size of an eyelash that becomes a brilliant bouquet of zinnias, or transforming a shedful of debris into an efficiently humming produce wash-and-pack house. There are so many tangible rewards to be had in farming. The cherry on top of this rich, delicious cake of satisfaction is that you can create objects and systems that are absolutely unique and even aesthetically pleasing to your eye. You have the freedom to paint the barn purple should you wish, or to install homemade sculptures alongside the farm road. The farm is your own four-dimensional (don't forget about time!) canvas upon which you can paint your future. What form the final expression of any impulse takes is unpredictable, but that you get to exercise creative license is a benefit that most people don't get in their jobs.

SO MANY RICHES

Of all the ways there are to make a living, farming is a special and strange path. It's a lifestyle choice on top of a career path. Farming is a risky enterprise—financially, physically, and emotionally. Your lifestyle paycheck is akin to a high-yielding reward on a crazy new tech stock. And, just like stock investing, the risky bets yield the highest rewards. Still, your highest returns on investment will involve intangibles—time with your family, freedom of creative expression, growing what you love, and having fun. These are things that many people can only dream of; but for farmers, they come with the job description. Your ability to self-determine, to work with and in nature, to grow much-needed nutrition for your fellow citizens—these are the unheralded currencies of our agricultural banking system. Enjoy them mightily!

CHAPTER REVIEW QUESTIONS

1. What rewards beyond money do you hope to achieve with your business?

2. How will your family and personal life intersect with your business? What good will the farm bring to the people you love?

3. How will the farm positively (or possibly negatively) impact your health and well-being? How will you mitigate any negative impacts?

4. What aspects of the farm will bring you joy? How can you keep that joy front and center for yourself and the people whom your business touches?

We've never been more excited about the potential of a career in farming than *right now*. It's taken decades to reach the point where market conditions are this perfect for small to midsize farms. Consumer demand, customer awareness, technological advancements, and social media have all intersected to tip the scales toward the independent producer. The horizon is ripe for enduring prosperity. If you can dream it—and then grow it—chances are nearly certain that there's a market for it.

The rules are changing in real time. The fact that you can now reach ten thousand self-identifying customers with the click of a button is mind-boggling, unimaginable when we first began farming. For example, the simple addition of Forrest's farm store to Google Maps immediately boosted monthly sales by 25 percent. Maintaining social media is just as easy. For five minutes of your time, you can shoot and edit a video, upload it onto an app, and—ta-da!—the world is watching you farm. This is an opportunity to share your stories and forge enduring connections.

Advancements in packaging, branding, storage, refrigeration, and transport have been amazing, as well. Early on, we often felt like we had to reinvent the wheel, from freezers to vacuum packers to labeling machines. Today, just try to keep up with the deluge of solutions that are constantly debuting. At a recent farmers' market in Washington, DC, a robot rolled past on its way to deliver pizzas! By the time you read this, you might be receiving home deliveries by drone, the sky abuzz with mechanical bees.

A day is coming, too, when consumers will have unprecedented access to what our farms look like. There are already airborne drones videoing of some of our country's most disturbing CAFOs (concentrated animal feeding operations), revealing in real time how certain food factories pollute the land, air, and water around them. This is great news for sustainable and biological farmers alike. "Come on over and take a look around," we can say with confidence. Our farms will offer full transparency, alive with healthy plants and animals, with its human stewards happily tending to the work.

Similarly, the unstoppable food-safety movement will bring ever more scrutiny to agriculture. We haven't yet addressed FSMA, the FDA Food Safety Modernization Act, but it's an important part of the future of our food landscape. Now, both consumers and the government want to know how clean and safe our farms are for producing the nation's food supply. While terms such as *clean* and *safe* come with nuanced definitions worthy of continued, educated debate, we want to stay precisely on the "good guy" side of the equation. Keeping our farms tidy and organized, implementing sensible food-safety procedures, and practicing timely, accurate record keeping prove that we take our role as food producers seriously.

For those of you truly starting from scratch, this means that you have no bad habits to break! Your awesome opportunity to get started on the right foot, with eyes wide open, begins now. For those of you already into your farm journey, make sure that you cover these bases of transparency and organization. Take pride in being a producer of safe, reliable, nutritious food. The resources already exist to assist you in this goal.

Technology will keep advancing, and this is precisely why we're so excited for people who want to get their hands in the dirt. Progress has always been led by those willing to invest sweat equity and to do the jobs that no one else has wanted to do. Now, just like millions of programmers, engineers, designers, and IT specialists, farmers must get paid for their work, too. But, it will only happen if you remain insistent and personally make it happen. The opportunity is right before you.

That's the goal, ultimately: to adapt your farming drive and passion to the twenty-first century, however it unfolds. Adjusting to real-time lessons and ever-changing market conditions will always be a huge component of your success. Honoring old-fashioned wisdom and sustainable methods, you can still farm much like previous generations did. But, to make a profit in agriculture, you can no longer run your farm business the same way.

YOUR FIRST DAY

What will your first day as a farmer be like? Whatever your circumstances happen to be—intern, apprentice, lessee, or owner—when it comes to finally lacing up your boots and stepping into your own fields for the first time, everyone truly starts on level ground. It makes little difference if you grew up on a farm, or if you're just experiencing productive soil for the first time. You might be in California, or you might be in New York; you might be staking tomatoes, building a fence, or clearing a field. It doesn't especially matter what you're doing, or where you happen to be. Regardless of the particulars, the emotions will essentially be the same.

You'll have butterflies. You'll wonder where to begin, how to start, and what to do first. You'll be overwhelmed by the sheer enormity of your responsibility. You'll be so excited to get going, and you'll probably try to get everything done on the very first morning.

Forrest once had an apprentice who repeatedly tripped over his own feet pulling brush his first day, overeager to make a good impression. What he thought was a sprint was actually an endurance event. By noon, he was exhausted, but still in good spirits. Over the course of his one-year apprenticeship, he gradually developed a sustainable pace.

Ellen has probably had 250 first-timers on her farm over the years. They were all so clean and shiny on the first day! And, inevitably, by 5:00 PM, no matter the time of year, they had dirt on their cheeks, ragged fingernails, and a big smile plastered on their faces. Oh, and a rosy blush of sunburn wherever skin was exposed. Day two usually began with an inventory of which muscles were newly awakened, and where the scratches were. No matter. Every single day, they would get better at farming. And, so will you.

Take a deep breath. Successful farmers have all been there before, and they've come through it. Ellen remembers early on how, each time she approached a new task, she would suddenly become a person who talked to herself out loud. Her main refrain was, "Why is this so hard?" Then, after some head-scratching, multiple trips to the barn, and a few additional mess-ups, she'd have a revelation. "Ohhhhh," she'd say, finally solving the problem. "That's better."

Whether audibly or not, make sure that you self-talk your way through both the positive and negative emotions. Give yourself a pat on the back, at least in your mind's eye, as, hour by hour, you figure things out.

YOUR FIRST HARVEST

We wish we could see the look on your face when you harvest your first strawberries, pick your first cherry tomatoes, or discover that, sometime in the middle of the night, your sow has given birth to fourteen healthy, hungry, wriggling piglets. Because it's gonna be beautiful.

Ellen's first harvest was near Davis, California, in 1987, on a little patch of ground rented from a wily old farmer and jack-of-all-trades. He loved doing the field preparation with his D5 Caterpillar, a massive, track-type bulldozer—not your average farm tractor! Her first—and as it turned out, last—crop at this farm was beets, fifteen hundred row-feet planted like corn, each row thirty inches apart. The soil was a lousy, sticky clay, but those beets still grew. And they turned out to be gorgeous, perfectly shaped, gleaming ruby orbs. With a rubber band around their stems and a good rinse, they were packed into a few stray boxes, and off to the Davis Food Co-op they went. She was overwhelmed with pride. To get all that product from a single cup of seed was absolutely stunning.

Five-year-old Forrest was his grandmother's finest employee, dutifully carrying an empty tin pail to her chicken coop each afternoon to gather the day's eggs. There, in a cool, cinderblock building, shaded by walnut trees, he'd carefully reach beneath the hens to explore the dark, downy depths. Eureka! A warm brown egg, or sometimes two. Into the pail they went, gently. His grandmother would transform these golden yolks into golden pound cakes, her regional culinary claim to fame. He'd help her with these, too, selling them after church from the trunk of her 1973 dark green Pontiac.

Harvest—and subsequently, sales—came early for both of us. You might not have a D5 Caterpillar, or a cake-baking dynamo of a grandma, but your own experience will be no less memorable. This is one of the biggest secrets about farming: the pure joy of growing your own food, and having your first sale.

YOUR FIRST SALE

At long last, you've produced something with your own hands, cultivating it from field to harvest. It's a product you practically dreamed into existence. Now, you've found a customer who sees the value in your hard work and wants to buy what you have. What an amazing feeling! How many people get the chance to create what they love, and get paid for it? This is rare territory.

Even after decades of farming, it remains an experience neither of us takes for granted. Ellen's first market season was at age sixteen, selling produce she had helped grow at Potomac Vegetable Farms. After only a couple of weeks of training, she was sent off to the Fairfax farmers' market alone, driving an ancient F-150 with a "three-on-the-tree" manual transmission. This was so long ago in farmers' market history, that the farm used hanging scales and homemade, heavy wooden tables, with holes cut in the center to thread a beach umbrella through. Much of the produce simply sat in baskets on the ground, for folks to root around in.

But good golly. Taking in those one-dollar bills and making the customers smile was incredibly gratifying. The loop between seed and the money to buy the next seed was closed. Arriving back at the farm with empty baskets but an apronful of money topped off the whole experience. This kind of sales job made all kinds of sense—she was selling the literal fruits of her labor to appreciative eaters.

ENTROPY AND RENEWAL

By definition, agriculture does not exist without us, the farmers. But what control do we truly have? Our kindly and wise observance of the laws of nature—our obeisance—serves us well every day. But our job as producers is not only to work with nature, but occasionally to protect ourselves from it, as well.

It's not rainbows and sweet breezes all the time. When a fluke killing frost is forecast for June, you'd better get out some protective row cover. When fifty-mile-an-hour winds threaten to blow your greenhouses clear to the next county, you'd better batten down the hatches. If a herd of twenty deer find your sweet-potato patch, it will be toast, so you'd better start setting some tall fence posts.

Entropy is part of nature—the inexorable movement into disorder. It takes intention and effort to keep chaos from overtaking your farm. Therefore, it's your job to strike a thoughtful balance between reverence and resistance. Honor natural cycles, reciprocity, and diversity by designing systems that foster growth of soil biology, all the while sustaining your economic bottom line. But nature alone will not care for your crops; it's up to you to nurture them. So make sure to water delicate young plants when it doesn't rain. Provide shelter from howling winds when your ewe goes into early labor during a late March snowstorm.

It starts early, doesn't it, our respect for nature? Maybe it began in your mother's flower garden, or with a windowbox filled with herbs. Or perhaps even picking dandelions on the school playground, staring in wonder at the complexity of a tiny yellow sun cupped in your hands. As other kids dashed past, playing tag, you sat dumbstruck, marveling. Then, showing off, you ate that dandelion right in front of your friends. It was bitter, but nutritious!

At some point, regardless of our backgrounds, we all feel this moment of amazement, this unexpected awe and respect for nature. Gradually, as you spend more time working outdoors, as the soil and the sun become part of your daily life, you come to realize that you have so much more to offer to the world than you're commonly told or taught. And if you're lucky—and brave enough—you someday get the chance to follow this passion, and do something extraordinary.

DO SOMETHING EXTRAORDINARY

Aldo Leopold, in his seminal 1940s farming book, *A Sand County Almanac*, astutely observed how our culture often values businesses that make money at any cost—what he termed a "wrong action"—more than those that take "right actions" but generate a smaller profit. Back in Leopold's day, fighting for environmental responsibility, sustainable agriculture, and ethical for-profit businesses was considered bold and extremely progressive. This farmer had the temerity to speak out against corporate greed, and speak up on behalf of nature, which has no human voice.

But today, you have an even greater opportunity. What you've been exploring in this book is the idea of a complicated, dynamic, and holistic approach to agriculture. Your calling, then, is to take Leopold's observation and build upon it.

What if it were possible to perform a right action that pays? We contend that the day has arrived when you can live your values while also making robust, sustainable profits. It's a wonderful example of triple bottom-line thinking: Take care of your economic well-being, while at the same time treating your workers, customers, and neighbors with respect. All the while, you create a net-positive ecological impact.

We all learn from the masters, and Leopold set this stage more than a half-century ago. Now you have the chance to do something that he might never have even imagined. Pulling this off on a larger cultural scale would be extraordinary.

START YOUR FARM

The word for the soulfulness of a place is *numen*. This is the life force of your farm. The rules for maintaining this soulfulness are extremely straightforward: Give the soil what it needs, and the numen grows; take and take, and you'll feel it fading away. As the land's steward, this requires your intentionality, the wisdom to regard your farm as a holistic, living spirit, not just some biological resource. In our experience, the more we give, the more we can feel the land giving back. Consider these lines from Walt Whitman's "Song of Myself":

> Have you reckoned a thousand acres much? Have you reckoned the
>   earth much?
> Have you practiced so long to learn to read?
> Have you felt so proud to get at the meaning of poems?
> Stop this day and night with me and you shall possess the origin of all
>   poems,
> You shall possess the good of the earth and sun . . . .

When you "read" your landscape, study its hills and dells, its fields and horizons, what does it say to you? Surely it speaks. Thinking back to when we were new farmers, we wanted all of our questions answered, and right away, too! But we eventually learned the pragmatism of allowing space for some mystery on the farm.

This is the mystery of vernal frogs singing by moonlight, while the snow still blankets the ground. It's the rush of sap into the fruit trees, limbs russet and sweet, spring robins bowing the branches. It's the first wobbly steps of a newborn calf as its mother licks it clean, able to walk a mile before the sun dips low. It's the final embers in the evening hearth, cloaked in smokey ash, yet kindled to life the next morning with a nudge of hickory.

Sunflowers turn to face the dawn. A newly hatched chick heads straight to water. We forget that life is intuitive. We don't know exactly why, or how, we do much of what we do ourselves. But we know when it feels right, and we know when it feels good.

It's as easy to explain these things as it is to explain why, exactly, we are called to be farmers. As caregivers, life-bringers, and volunteers, we accentuate the good of the earth and the sun, nurturing plants from the soil for the benefit of humankind. Have we reckoned this much? What a responsibility. What an honor!

We place a seed into fertile ground, water it, and tend to it. The sun arches high overhead, and a delicate green shoot rises in response, touching the light. Why try to explain it? The only thing we know for sure is that when we pour our passion into what we love, we end up with more than we give. It's not necessary to fully understand. Some secrets are best left to the soil, hiding in plain sight beneath our feet.

ACKNOWLEDGMENTS

FORREST

I have so many people to thank for making this book possible. At the top of the list are my editor, Anna Bliss, and my coauthor, Ellen Polishuk.

It's been a joy to work again with Anna, and doubly so on such a different project than when we first collaborated. I've learned so much from her over the years—beyond her gift for clear-minded editing, she's taught me, a semi-feral farmer, how to behave like a professional in the literary world. Her attention to detail and willingness to address every kind of question have been a tremendous gift and helped to make the entire process seem practically effortless. Anna, thank you so much for your time and your talents.

Ellen, in several ways, this book couldn't have been possible without you. Two decades ago, at a farming conference, you nudged me toward economic sustainability when you insisted I apply for a farmers' market in the "big city." I was thoroughly intimidated, but I followed what turned out to be your excellent advice! Then, seventeen years later—again, at a farming conference—I figured I'd return the favor and challenge you, as well. I hope you'll agree that the results have exceeded both of our expectations. I look forward to remaining your friend and continuing this inexplicable farming journey together.

Many thanks are due to Stephany Evans, who worked to shape this book in its early stages, as well as to Matthew Lore, for his patience and goodwill as the project took shape. Matthew, thanks for giving this book an excellent home at The Experiment. Thanks to art director Sarah Smith and the design team for all your assistance with creating a beautiful book inside and out; and to Jennifer Hergenroeder, Ashley Yepsen, and Elizabeth Johnson for helping share this project with the rest of the world.

To Molly M. Peterson, I can't overstate what an honor it is to have your photographs on the front cover. You are such a talent, and I'm grateful to have you as my friend. Good luck to you and Mike in your new endeavors at the Stone Barns.

I'd like to thank my fellow farmers for their advice and feedback on this project and for fielding all of my oddball questions along the way. In no order, special thanks to Lyle Tabb, Steve Ernst, Mark Toigo, Don Magnani, Eric Plaksin, Rachel Bynum, David Weiss, and Chester Beahm for always fielding my inquires with an attitude of earnest helpfulness.

To Chris Nolte, who all those years ago politely insisted that I treat farming like any other business, thank you. Large parts of this book wouldn't exist without you and your no-nonsense business acumen.

To my son, Linus, who wants to be a shepherd and who made it so easy to be a part-time dad while I wrote, thank you. I'm very proud of you, kid.

Finally, in memoriam, I'd like to thank my farming grandparents, who, through decades of hard work and saving, made it possible for each of their grandchildren to attend college, debt-free. I am so blessed by the love you showed, and I hope in some small way this book honors and upholds the spirit that you passed along to me.

ELLEN

Forrest Pritchard came to me with the idea of *Start Your Farm* and asked me if I'd like to cowrite it with him. Without hesitation, I eagerly agreed. We brainstormed the chapters and concepts, and off we went from there. This is my first book-writing journey, and what a pleasure it has been to work alongside Forrest, riding his coattails. He was doggedly patient, consistently inspiring, and unflaggingly kind throughout. Thank you, Forrest, for inviting me to participate.

My second most constant companion in this process has been Anna Bliss, our editor. She was tolerant of my ignorance, and ever so gentle, yet firm, as she guided me to increasing clarity in the manuscript. Thank you, Anna, for being a constant believer in the mission, and for treating me with such good care.

Many thanks go to my technical advisors, Allen Philo and Jim Munsch. Allen lent his wisdom to the soils chapter. Thanks, Allen, for continuing to inspire me to love soil more and more, and for constantly expanding my curiosity and understanding of science. Jim and I have worked many hours together teaching farmers around the country how to use Veggie Compass. He is so smart, very funny, and excellent at communicating complex topics. Thanks, Jim, for helping me with the farm economics language, and for continuing to teach me how to look at and think about farm businesses.

Of course, none of this would have come about at all had my mother, Mary Coyne, not let me dig around in the dirt as a kid. She helped me get my first garden plot at age ten. I'll never forget how your understanding and appreciation of my farmer role came fully into being when you came to babysit Aaron every morning those first few months. Thanks, Mom, for allowing my farm dream to grow.

Aaron, my one and only boy. Your spirit and unwavering love astound me. Keep on shining your amazing light in this world.

And, last but not least, I thank Daniel Leggett, my partner and husband. Thank you for believing in me every day, for cheering me on, and for loving me so well. Your unconditional support continues to exceed my wildest hopes.

RESOURCES AND RECOMMENDED READING

FARMING WEBSITES AND PODCASTS

Ag America Lending (private lender specializing in farm loans of all sizes): agamerica.com

ATTRA (technical growing information for all enterprises): attra.ncat.org

The Carrot Project (promotes financial skills for farmers): thecarrotproject.org

Civil Eats (articles relevant to food and agriculture): civileats.com

Dirt Capital Partners (investor in farmland in the Northeast, promoting sustainable farming access and security): dirtpartners.com

Farm Commons (legal information for growers): farmcommons.org

Farm Service Agency (a USDA agency that focuses on farm conservation and regulations): fsa.usda.gov

Farmer to Farmer Podcast (one-on-one interviews with market growers): farmertofarmerpodcast.com

International Farm Transition Network (a network of professional service providers who support farm businesses during transitions and successions): farmtransition.org

Iroquois Valley Farms (a food and farmland company that invests in acreage for organic farmers): iroquoisvalley.com

Kickstarter (a platform for crowdsourcing capital from the community): kickstarter.com

The Land Connection (trains farmers in restorative farming techniques and promotes public education about sustainable farming): thelandconnection.org

The Land Institute (research on perennial grain and seed crops and ecologically intensified polycultures): landinstitute.org

Land Stewardship Project (promotes sustainable agriculture and healthy communities): landstewardshipproject.org

National Young Farmers Coalition (a national network of young and sustainable farmers to promote policy and organize like-minded farmers): youngfarmers.org

The Counter (articles relating to food, agriculture, and society): thecounter.org

On Pasture (weekly blog featuring articles about sustainable livestock practices): onpasture.com

Renewing the Countryside (champions farmers and businesses who renew the countryside through sustainable practices): renewingthecountryside.org

The Savory Institute (restoring decertified grasslands via holistic management): savory.global

Shared Earth (funder that promotes habitat restoration for broad biodiversity): sharedearth.com

Veggie Compass (free Excel spreadsheet and farm record-keeping information): veggiecompass.com

Vilicus Capital Partners (links socially minded investors with organic farmers): vilicusventures.com

BOOKS

Farming

Berry, Wendell. *The Unsettling of America: Culture and Agriculture*. Random House, Ex-library edition (June 12, 1982).

Clark, Andy. *Managing Cover Crops Profitably*. 3rd ed. College Park, MD: SARE Outreach, 2007.

Coleman, Eliot. *The New Organic Grower: A Master's Manual of Tools and Techniques for the Home and Market Gardener*. White River Junction, VT: Chelsea Green Publishing, 1995.

Fortier, Jean-Martin. *The Market Gardener: A Successful Grower's Handbook for Small-Scale Organic Farming*. Gabriola Island, BC: New Society Publishers, 2014.

Fukuoka, Masanobu. *The One-Straw Revolution: An Introduction to Natural Farming*. New York: New York Review Books, 2009.

Hartman, Ben. *The Lean Farm: How to Minimize Waste, Increase Efficiency, and Maximize Value and Profits with Less Work*. White River Junction, VT: Chelsea Green Publishing, 2015.

Leopold, Aldo. *A Sand County Almanac*. New York: Oxford University Press, 1949.

Magdoff, Fred, and Harold van Es. *Building Better Soils for Better Crops*. 3rd ed. College Park, MD: SARE Outreach, 2010.

Mefferd, Andrew. *The Greenhouse and Hoophouse Grower's Handbook: Organic Vegetable Production Using Protected Culture*. White River Junction, VT: Chelsea Green Publishing, 2017.

Salatin, Joel. *You Can Farm: The Entrepreneur's Guide to Start and Succeed in a Farming Enterprise*. Swoope, VA: Polyface, 1998.

Shepard, Mark. *Restoration Agriculture: Real-World Permaculture for Farmers*. Greeley, CO: Acres U.S.A., 2013.

Thistlewaite, Rebecca. *Farms with a Future: Creating and Growing a Sustainable Farm Business*. White River Junction, VT: Chelsea Green Publishing, 2013.

Zimmer, Gary, and Leilani Zimmer-Durand. *Advancing Biological Farming*. Greeley, CO: Acres U.S.A., 2011.

Zimmer, Gary, and Leilani Zimmer-Durand. *The Biological Farmer: A Complete Guide to the Sustainable and Profitable Biological System of Farming*. 2nd ed. Greeley, CO: Acres U.S.A., 2016.

Finance and Business

Allen, Robert G. *Multiple Streams of Income: How to Generate a Lifetime of Unlimited Wealth!* 2nd ed. Hoboken, NJ: Wiley, 2005.

Gerber, Michael E. *The E-Myth Revisited: Why Most Small Businesses Don't Work and What to Do About It*. New York: HarperCollins, 1995.

Hayes, Kathryn, Judith Gillan, Eric Toensmeier, Michelle Wiggins. *Exploring the Small Farm Dream: Is Starting an Agricultural Business Right for You?* 2nd ed. Belchertown, MA: The New England Small Farm Institute, 2009.

Hill, Napoleon. *Think and Grow Rich*. 1st ed. (1937). Shippensburg, PA: Sound Wisdom, 2016.

Kiyosaki, Robert T. *Rich Dad Poor Dad: What the Rich Teach Their Kids About Money—That the Poor and Middle Class Do Not!* Scottsdale, AZ: Plata Publishing, 2011.

Padgham, Jody, Craig Chase, and Paul Dietmann. *Fearless Farm Finances: Farm Financial Management Demystified*. 2nd ed. Spring Valley, WI: Midwest Organic and Sustainable Education Service, 2017.

Shanks, Julia. *The Farmer's Office: Tools, Tips and Templates to Successfully Manage a Growing Farm Business.* Gabriola Island, BC: New Society Publishers, 2016.

Wiswall, Richard. *The Organic Farmer's Business Handbook: A Complete Guide to Managing Finances, Crops, and Staff—and Making a Profit.* White River Junction, VT: Chelsea Green Publishing, 2009.

MAGAZINES AND ARTICLES

Acres U.S.A.: acresusa.com

American Vegetable Grower: growingproduce.com

Growing for Market: growingformarket.com

Longstroth, Mark. "Lowering the Soil pH with Sulfur." Michigan State University Extension. canr.msu.edu/uploads/files/Lowering_Soil_pH_with_Sulfur.pdf.

Mother Earth News: motherearthnews.com

Nordell, Anne and Eric. "Weed the Soil, Not the Crop." *Acres U.S.A.* 40, no. 6 (June 2009). organicfarmingworks.com/wp-content/uploads/June09_Nordells.pdf.

The Stockman Grass Farmer: stockmangrassfarmer.com

Vossen, Paul. "Changing pH in Soil." University of California, Cooperative Extension. vric.ucdavis.edu/pdf/Soil/ChangingpHinSoil.pdf.

INDEX

Page numbers in *italics* refer to figures, tables and charts.

E

earnings, reporting, 153
ecological stewardship, 142, *142*, 235
economics
  about, 139–40
  accounting and bookkeeping, 151–54
  depreciation, 154–55
  economic forces, uncontrollable, 139–40
  lifestyle paycheck, 223
  metrics, proprietary, 139
  prices, setting, 143–51, *148–49*
  profit, 140–43, *142*
  record keeping, 155–57
  sustainability, as concept, 16–17
  in triple bottom line approach, 142, *142*, 235
  See also finances
education
  about, 52–53
  beginner farmer training programs, 60
  farm incubators, 60–61
  farm tours, DIY, 63
  formal, 9, 53
  going it alone, 53–54
  internships and apprenticeships, 57–58, 192–93
  mentors, 55, 56–57, 77–79
  need for, 52
  off-farm educational avenues, 61–63
  on-farm learning models, 57–61, 192–93
  post-college farm work, 55–57
  stay-at-home, 63–64

summer jobs, 54, 55
wages, working for, 58–60
effluent, 96
eggs, 149–50
employer and employee relationship
  compensation, 191–92
  housing, 192
  insurance, 192
  legal responsibilities, 191
energy
  personal, 15–16
  renewable, 113
Enron, 8
enterprise budgets, 145–46, 147
enterprise failures, 208–9
entropy and renewal, 233–34
envelope system for budgeting, 167–68
environmentalism, 15
equipment
  about, 45–46
  borrowing, 46
  buying outright, 46
  doing without, 47
  failures, 212
  hiring custom operator, 46
  issues with, 176
  outsourcing, 46
  record keeping for, 156
  renting, 46
Ernst, Steve, 125–26
error, human, 205
Evans, Bob, 21
exercise, 220–21
expansion failures, 209–10
expenses
  daily, 159–60, 164–65

recurring, 165
experience
  accounting, 24
  business, 23–24
  importance of, 17
extension agents, 62, 77, 95–96

F

Fabaceae (legumes), 37, 99
failure
  about, 203–6
  chickens, attrition of, 206–7, 208
  cost of avoiding, 207
  enterprise, 208–9
  equipment, 212
  expansion, 209–10
  of faith, 215–16
  human error, 205
  importance of, 18
  life-threatening, 214–15
  nature, lack of control over, 203–5
  seasons, working with, 208
  small, 213–16
  staffing, 213–14
  systems, 211–12
faith, failures of, 215–16
family
  farm parents/kids, 54–55
  lifestyle paycheck, 219–20
  loans from, 31
  relationship with, 197–200
  as unpaid workers, 190
Farm Credit operating loans, 154
Farm Crisis, 1980s, 73–74
farm economics. *See* economics

farm education. *See* education
farmers
  age, average, 55
  income, *32, 176*, 177
  motivation for becoming, 11–12
  percentage of Americans as, 7, *13*, 14
  perceptions of, 10
  relationships with other, 196–97
  as volunteers, 10–11
farmers' markets, 111–13, *112*, 119–20
farm incubators, 60–61
farming
  college classes on, 9
  as college graduate career choice, 8–9
  expectations, redefining, 21
  opportunities in, *13*, 14
  scrutiny of, 230
  twenty-first century, *75*, 75–76
  See also specific topics
farmland
  about, 66–69
  accessing, solo, 79–81
  accessing, with mentorship, 77–79
  affordability issues, 3–4, 66–67, *72*, 72–75, *73*
  amount of, *13*, 14
  buying, 82–86, *84*
  cheap food, consequences of, 69–71
  cost, accounting for, 67–69
  farming without, 81
  free, 71–72, 80–81
  hilly, 87
  real estate values, *72*, 72–73
  renting/leasing versus buying, 79–80
  sloped, 87
  transfer of, *75*, 76

S

ABOUT THE AUTHORS

FORREST PRITCHARD is the *New York Times*–bestselling author of *Gaining Ground: A Story of Farmers' Markets, Local Food, and Saving the Family Farm*, as well as *Growing Tomorrow: Behind the Scenes with 18 Extraordinary Sustainable Farmers Who Are Changing the Way We Eat*. A graduate of the College of William & Mary with degrees in English and geology, Pritchard is also a full-time organic livestock farmer and seventh-generation producer. His books have been named top reads by NPR, *Washington Post, Los Angeles Times, Publishers Weekly, Library Journal*, and many more.

ELLEN POLISHUK is a first-generation sustainable vegetable farmer, holding a degree in horticulture from Virginia Tech. A self-described "compost queen," Ellen grew up in the suburbs of Washington DC in the 1960s and 70s. In 1992, with five farm seasons under her belt, she was hired by Potomac Vegetable Farms to manage their satellite farm in Virginia. Formerly an owner of Potomac Vegetable Farms, Ellen is now a sought-after farm consultant and conference speaker, and a passionate advocate for the business of farming.